T0113917

Criminal Drone Evolution: Cartel Weaponization of Aerial IEDs

Criminal Drone Evolution: Cartel Weaponization of Aerial IEDs

Robert J. Bunker and
John P. Sullivan, Editors

A Small Wars Journal–El Centro Anthology

Front Cover Image: Toma Anguilla CJNG (CJNG Takes Aguillia). 4 July 2021, AGUILLA, Michoacán de Ocampo – The municipality of Anguilla, located 270 kilometers southwest of Morelia, Michoacán in Mexico's Tierra Caliente was taken over by elements of the *Cártel Jalisco Nueva Generación* (CJNG) on 5 April 2021. Here we see an CJNG improvised armored vehicle (IAFV) supported by a DJI Mavic 2 Zoom aerial drone (sUAV). The IAFV has seen combat, as evidenced by the bullet holes in the hood, and contains a turreted .50 cal. M2 Browning machine gun (or foreign copy). The IAFV crew are members of the *Fuerzas Especiales Mencho* (FEM), the CJNG special forces unit. Despite the deployment of *Guardia Nacional* and SEDENA forces, the armed conflict between the CJNG and the *Cárteles Unidos* (United Cartels) continues. Photo: "Toma Aquillia CJNG," 4 July 2021 © Cuartoscuro, used under license.

Rev. date: 10/22/2021

CONTENTS

    Why Drones – Why Now?

    David Hambling

    Weaponized Drones and Order-Of-Battle (OOB) Criteria

    Lisa J. Campbell

    Criminal Armed Groups and Drones – from
    Smuggling and ISR, to Improvised Weapons Platforms

    John P. Sullivan and Robert J. Bunker

    Mexico's Cartels Building Custom-Made Narco Drones: DEA

    Camilo Mejia Giraldo

    Mexican Cartel Tactical Note #21:
    Cartel Unmanned Aerial Vehicles (UAVs)

    Robert J. Bunker

About Small Wars Journal
and Foundation

Small Wars Journal facilitates the exchange of information among practitioners, thought leaders, and students of Small Wars, in order to advance knowledge and capabilities in the field. We hope this, in turn, advances the practice and effectiveness of those forces prosecuting Small Wars in the interest of self-determination, freedom, and prosperity for the population in the area of operations.

We believe that Small Wars are an enduring feature of modern politics. We do not believe that true effectiveness in Small Wars is a 'lesser included capability' of a force tailored for major theater war. And we *never* believed that 'bypass built-up areas' was a tenable position warranting the doctrinal primacy it has held for too long—this site is an evolution of the MOUT Homepage, Urban Operations Journal, and urbanoperations.com, all formerly run by the *Small Wars Journal's* founding Editor-in-Chief.

The characteristics of Small Wars have evolved since the Banana Wars and Gunboat Diplomacy. War is never purely military, but today's Small Wars are even less pure with the greater inter-connectedness of the 21st century. Their conduct typically involves the projection and employment of the full spectrum of national and coalition power by a broad community of practitioners. The military is still generally the biggest part of the pack, but there are a lot of other wolves. The strength of the pack is the wolf, and the strength of the wolf is the pack.

The *Small Wars Journal's* founders come from the Marine Corps. Like Marines deserve to be, we are very proud of this; we are also conscious and cautious of it. This site seeks to transcend any viewpoint that is single service, and any that is purely military or naively U.S.-centric. We pursue a comprehensive approach to Small Wars, integrating the full joint, allied, and coalition military with their governments' federal or national agencies, non-governmental agencies, and private organizations. Small Wars are big undertakings, demanding a coordinated effort from a huge community of interest.

We thank our contributors for sharing their knowledge and experience, and hope you will continue to join us as we build a resource for our community of interest to engage in a professional dialog on this painfully relevant topic. Share your thoughts, ideas, successes, and mistakes; make us all stronger.

"...I know it when I see it."

"Small Wars" is an imperfect term used to describe a broad spectrum of spirited continuation of politics by other means, falling somewhere in the middle bit of the continuum between feisty diplomatic words and global thermonuclear war. The *Small Wars Journal* embraces that imperfection.

Just as friendly fire isn't, there isn't necessarily anything small about a Small War.

The term "Small War" either encompasses or overlaps with a number of familiar terms such as counterinsurgency, foreign internal defense, support and stability operations, peacemaking, peacekeeping, and many flavors of intervention. Operations such as noncombatant evacuation, disaster relief, and humanitarian assistance will often either be a part of a Small War, or have a Small Wars feel to them. Small Wars involve a wide spectrum of specialized tactical, technical, social, and cultural skills and expertise, requiring great ingenuity from their practitioners. The *Small Wars Manual* (a wonderful resource, unfortunately more often referred to than read) notes that:

> *Small Wars demand the highest type of leadership directed by intelligence, resourcefulness, and ingenuity. Small Wars are conceived in uncertainty, are conducted often with precarious responsibility and doubtful authority, under indeterminate orders lacking specific instructions.*

The "three block war" construct employed by General Krulak is exceptionally useful in describing the tactical and operational challenges of a Small War and of many urban operations. Its only shortcoming is that is so useful that it is often mistaken as a definition or as a type of operation.

We'd like to deploy a primer on Small Wars that provides more depth than this brief section. Your suggestions and contributions of content are welcome.

Who Are Those Guys?

Small Wars Journal is NOT a government, official, or big corporate site. It is run by Small Wars Foundation, a non-profit corporation, for the benefit of the Small Wars community of interest. The site was founded by Dave Dilegge, its inaugural Editor-in-Chief. Its current principals are David S. Maxwell (Editor-in-Chief) and Bill Nagle (Publisher), and it would not be possible without the support of myriad volunteers as well as authors who care about this field and contribute their original works to the community. We do this in our spare time, because we want to. McDonald's pays more. But we'd rather work to advance our noble profession than watch TV, try to super-size your order, or interest you in a delicious hot apple pie. If and when you're not flipping burgers, please join us.

About El Centro

El Centro is *SWJ's* focus on small wars in Latin America. The elephant in the hemispheric room is clearly the epidemic criminal, cartel, and gang threat, fueled by a drug and migration economy, rising to the level of local and national criminal insurgencies and a significant U.S. national security risk. *El Centro* explores those and other issues across the U.S. Southern Border Zone, Mexico, the Caribbean, Central and South America to develop a better understanding of the national and regional challenges underlying past, present, and future small wars.

The *El Centro* Main section presents relevant *Small Wars Journal* articles and *SWJ Blog* posts. Other sections have a reading list and research links of relevant external works. We do link to some Spanish language resources and occasionally put up an article in both Spanish and English, but we are pretty much mainly operating in English. We look forward to being able to roll out *El Centro*, en Español, dentro de poco.

The *El Centro* Fellows are a group of professionals with expertise in and commitment to the region who support *SWJ's* approach to advancing our field and have generously agreed to join us in our *El Centro* endeavor. With their help and with continued development on our site's news and library sections, we look forward to providing more *El Centro*-relevant *SWJ* original material and more useful access to other important works and resources in the future.

El Centro Fellows

The *El Centro* Fellows have expertise in and commitment to Latin America, support *SWJ's* particular focus on the small wars in the region, and agree with *SWJ's* general approach to advancing discussion and awareness in the field through community dialog and publishing.

El Centro Associates are actively engaged in research or practice in the region and in transnational organized crime or insurgency. The Fellows have already made significant and distinguished contributions to the field through the course of their career. The Senior Fellows are Fellows that are central to producing *SWJ El Centro* and are very active in managing our work in this focus area.

Senior Fellows

Robert J. Bunker
John P. Sullivan

Fellows

Michael L. Burgoyne
Edgardo Buscaglia
Irina A. Chindea
Guadalupe Correa-Cabrera
José de Arimatéia da Cruz
Steven S. Dudley
Douglas Farah

Vanda Felbab-Brown
Luis Jorge Garay-Salamanca
Ioan Grillo
Gary J. Hale
Nathan P. Jones
Paul Rexton Kan
Robert Killebrew
Max G. Manwaring
Molly Molloy
Robert Muggah
Luz E. Nagle
Alexandra Phelan
Eduardo Salcedo-Albarán
Robert H. Scales
Teun Voeten

Associates

Pamela Ligouri Bunker
Alma Keshavarz
Daniel Weisz Argomedo

Interns

Anibal Serrano

Past Fellows

George W. Grayson
Graham H. Turbiville, Jr.

The views expressed in this anthology are those of the author(s) and do not necessarily reflect the official policy or position of the Department of the Army, the Department of Defense, the Federal Bureau of Investigation, the Department of Justice, or the U.S. Government, or any other U.S. armed service, intelligence or law enforcement agency, or local or state government.

Contributors

Editors

Dr. Robert J. Bunker is Director of Research and Analysis, C/O Futures, LLC, and a Senior Fellow with *Small Wars Journal–El Centro*. He is also an Instructor in the Safe Communities Institute (SCI) at the Sol Price School of Public Policy, University of Southern California. He holds university degrees in Political Science, Government, Social Science, Anthropology-Geography, Behavioral Science, and History and has undertaken hundreds of hours of counterterrorism and counternarcotics training. Past professional associations include Minerva Chair at the Strategic Studies Institute, US Army War College and Futurist in Residence, Training and Development Division, Behavioral Science Unit, Federal Bureau of Investigation Academy, Quantico. He has weaponized UAS/C-UAS research and field exercise experience since 2014 pertaining to non-state threat use. Dr. Bunker has well over 500 publications—including about 40 books as co-author, editor, and co-editor—and can be reached at docbunker@smallwarsjournal.com.

Dr. John P. Sullivan was a career police officer. He is an honorably retired lieutenant with the Los Angeles Sheriff's Department, specializing in emergency operations, transit policing, counterterrorism, and intelligence. He is currently an Instructor in the Safe Communities Institute (SCI) at the Sol Price School of Public Policy, University of Southern California, a Senior Fellow with *Small Wars Journal–El*

Centro, and Associate with C/O Futures, LLC. Sullivan received a lifetime achievement award from the National Fusion Center Association in November 2018 for his contributions to the national network of intelligence fusion centers. He completed the CREATE Executive Program in Counter-Terrorism at the University of Southern California and holds a Bachelor of Arts in Government from the College of William and Mary, a Master of Arts in Urban Affairs and Policy Analysis from the New School for Social Research, and a PhD from the Open University of Catalonia (Universitat Oberta de Catalunya). His doctoral thesis was "Mexico's Drug War: Cartels, Gangs, Sovereignty and the Network State." He can be reached at jpsullivan@smallwarsjournal.com.

Contributors

Parker Asmann graduated from DePaul University in Chicago with degrees in Journalism and Spanish, and a minor in Latin American studies. He was a freelance reporter for various publications—including *These Times, Jacobin*, The North American Congress on Latin America (NACLA) and the *Security Assistance Monitor*—before joining *InSight Crime* in 2017 as a staff writer. He is presently working on a Masters in International Peace and Conflict Resolution from American University.

Lt. Col. Lisa J. Campbell is an Intelligence Officer with the California Air National Guard. She holds a Bachelor of Special Studies in Geology from Cornell College and an MBA from the University of La Verne. She has deployed numerous times in the CENTCOM, SOUTHCOM and EUCOM Areas of Responsibility. Her career focus areas include aircraft and air base survival, operability in combat environments, operational studies of adversary capabilities, and counter-terrorism.

Dr. Scott Crino is the CEO and co-founder of Red Six Solutions, LLC, which is a company dedicated to delivering Red Teaming consulting and services to its customers. He served 21 years in the US Army as an Attack Aviation pilot and Operations Research/Systems Analysis officer to include deployments to multiple combat zones and held several command and staff positions. Prior to becoming CEO of Red Six, he served as President of Crino Consulting Group and as a Managing Director at Teneo Holdings. He has BS and MS degrees in Industrial Engineering from the University of Buffalo and Texas A&M, respectively, and a PhD in Systems Engineering from the University of Virginia.

Conrad "Andy" Dreby is a co-founder of Red Six Solutions where he is the Director of Red Teaming. Andy is a retired US Army armor officer. For the past ten years. He has managed a wide variety of red team projects for the Defense Department, Homeland Security and commercial clients. He is an expert in using Red Teaming techniques and approaches to find vulnerabilities and identify solutions for the company's clients. He has a Master's in Economics from the University of Oklahoma and a Masters in Professional Studies in Homeland and Information Security from Pennsylvania State University.

Brenda Fiegel is a Senior Intelligence Analyst and the Editor of the *Latin American Operational Environment Watch* at the Foreign Military Studies Office (FMSO) in Fort Leavenworth, Kansas. FMSO conducts open-source and foreign collaborative research, focusing on the foreign perspectives of understudied and unconsidered defense and security issues. Her specific research expertise includes "US/Mexico foreign relations," "US/Mexico border security threats," "Mexican and Central American violence/extremist groups to include drug cartels" and "Conflict resolution and peacekeeping in Mexico and Central America." She has lectured on these topics in professional military education settings, at Interagency Security Conferences, at Customs and Border Patrol Facilities, and at academic forums. She holds a BA in International/Global Studies and a BA and a MA in Spanish Language and Literature from University of Wisconsin-Milwaukee.

Camilo Mejia Giraldo is a freelance writer from Medellin, Colombia who previously worked as a reporter in Australia and Indonesia. He has written for *InSight Crime, Colombia Reports, World Politics Review,*

Mongabay, and the BBC. He is the co-founder of photojournalism collective VELA Colectivo.

David Hambling is a journalist, author, and consultant whose work focuses on unmanned systems and other advanced technology. He writes for *Aviation Week, New Scientist, The Economist, Popular Mechanics*, and other publications. He is based in London. His first book, *Weapons Grade* (2005), explored the military roots of consumer technology, from microwave ovens to GPS and the Internet. *Swarm Troopers: How Small Drones will Conquer the World* (2015) examined the disruptive military potential of small-unmanned aircraft. His latest book, *We: Robot* (2018), is a wide-ranging study of commercial, industrial, social, scientific, and military robots and their impact on society. He has acted as consultant on emerging military technologies for a variety of organizations in the defense and security sector.

Katie Jones is a researcher at *InSight Crime* who specializes in environmental crime and has a sustained interest in human rights, citizen security, socio-economic development, and education in the Americas. She holds a BA in History and Politics from University of Warwick and a diploma in Journalism and Communications from Sciences Po Paris Campus de Reims. She is fluent in Spanish and French.

Dr. Alma Keshavarz is an Associate with C/O Futures, LLC and an Associate with *Small Wars Journal–El Centro*. She received her PhD in Political Science at Claremont Graduate University. Her dissertation focused on hybrid warfare applied to the Islamic State, Russia, and Iran's Islamic Revolutionary Guard Corps. She previously earned a MA in political science at the same institution. She also holds an MPP from Pepperdine's School of Public Policy and a BA in Political Science and English from University of California, Davis. She has held various research intern and associate positions and has served as a graduate assistant at Pepperdine University. Her research interests include non-state actors, specifically Hezbollah, cyber security and warfare, and national security strategy with a regional focus on Middle East politics, specifically Iran, Lebanon, Yemen, and Syria. She has written a number of *SWJ* articles and has also co-published a number of the works for the Foreign Military Studies Office (FMSO), Fort Leavenworth, KS. She is fluent in Spanish and Farsi and is a past Non-resident Fellow in Terrorism and Security Studies at TRENDS Research & Advisory.

David A. Kuhn is an Associate with C/O Futures, LLC and presently the principal of VTAC Training Solutions. He is a subject matter expert in analysis, technical instruction, and terrorism response training related to stand-off weaponry (MANPADS, threat, interdiction, aircraft survivability, et. al), infantry weapons, small arms, IED/VBIEDs, WMD, and other threat and allied use technologies. He has been a senior consultant and author for Jane's Information Group on terrorism response; unconventional weapons response, military standoff weapons, and critical infrastructure protection; and has co- authored a number of Jane's Response Handbooks. He has developed comprehensive advanced training courses designed to provide law enforcement agencies and Special Response Teams with incident response protocols for terrorist/s and/or active shooter/s situations. He has taught these courses to more than 51 law enforcement agencies including police department Special Response Teams (SRT), EOD units, FBI, DOJ, US Secret Service, US Marshal's Service, TSA, US Coast Guard, and US Customs.

Chivis Martinez is the pseudonym of a female Mexican blogger and site principal at *Borderland Beat*. She has been reporting on the Mexican drug war since 2011 at this blog site.

Maria Alejandra Navarrete is former researcher and writer for *InSight Crime* and presently a technical writer for Management Systems International, Bogota, Colombia. She has a BA in Political Science and Government and minors in Journalism and Business Administration from Universidad de Los Andes and a MS in International Relations from The London School of Economics and Political Science (LSE). She has also worked as a political journalist for Grupo Semana in Bogota and as an intern with Amnesty International in London. She is fluent in Spanish and has a professional working proficiency in French.

Dr. James J. Torrence is an active duty US Army Signal Corps officer. He is a graduate of the United States Military Academy. He has a Doctorate in Strategic Security and multiple graduate degrees including an MS in Strategic Design & Management, an MS in Cybersecurity, and a Master of Military Art & Science. He has deployed twice to Afghanistan as a battalion communications officer and has served in various military leadership positions in the United States, Germany, Belgium, Korea, and Israel. He is the author of *Strongpoint Cyber Deterrence* (SWJ Pocket Book, 2020).

Acronyms

3D	Three Dimensional
AFO	Arellano-Félix Organization
AFP	Associated Foreign Press
AIC	Agencia de Investigación Criminal; Criminal Investigation Agency (Defunct)
A-IED	Aerial-Improvised Explosive Device
AMLO	Andrés Manuel López Obrador
BNC	Bayonet Neill–Concelman (RF Connector)
BRACNA	Brigada Especial contra el Narcotráfico; Special Brigade Against Drug Trafficking
C^2	Command & Control
C4	Composition C-4; Plastic Explosive
CAF	Cártel Arellano Félix; Arellano-Félix Cartel
CAG	Criminal Armed Group
CBP	Customs and Border Protection
CDS	Cártel de Sinaloa; Sinaloa Cartel
CENTCOM	Central Command (US)
CG	Center of Gravity
CIA	Central Intelligence Agency

CISEN	Centro de Investigación y Seguridad Nacional; Center for Investigation and National Security (Defunct)
CJNG	Cártel Jalisco Nueva Generación; Jalisco New Generation Cartel
CMI	Coordinación de Métodos de Investigación; Coordinator of Investigative Methods
COTS	Commercial Off-The-Shelf
COVID19MX	Coronavirus Disease 2019 Mexico
CSIS	Centro de Estudios Estratégicos Internacionales; Center for Strategic and International Studies
CSRL	Cártel de Santa Rosa de Lima; Santa Rosa de Lima Cartel
CTNG	Cártel de Tijuana Nueva Generación; Tijuana New Generation Cartel
C-UAS	Counter-Unmanned Aerial Systems
DEA	Drug Enforcement Administration
DHS	Department of Homeland Security
DIY	Do-It-Yourself
DJI	Da-Jiang Innovations
DSTL	Defence Science and Technology Laboratory
DTO	Drug Trafficking Organization
EU	European Union
EW	Electronic Warfare
FAA	Federal Aviation Administration
FARC	Fuerzas Armadas Revolucionarias de Colombia; Revolutionary Armed Forces of Colombia
FEM	Fuerzas Especiales Mencho; Mencho Special Forces
FFAA	Fuerzas Armadas; Armed Forces

FGR	Fiscalía General de la República; Attorney General of the Republic
FLIR	Forward Looking Infrared
FMSO	Foreign Military Studies Office
FOS	Frente Oliver Sinisterra; Oliver Sinisterra Front
FT	Feet
G7	Group of Seven
GHZ	Giga Hertz
GOM	Government of Mexico
GPS	Global Positioning System
GUP	Guerrillas Unidas del Pacífico; United Guerrillas of the Pacific
I&W	Indications & Warnings
IAFV	Improvised Armored Fighting Vehicle
ID	Identification
IED	Improvised Explosive Device
IoT	Internet of Things
IR	Infrared
IS	Islamic State
ISIS	Islamic State in Iraq and Syria / Islamic State of Iraq and al-Sham
ISR	Intelligence, Surveillance, and Reconnaissance
KM	Kilometer
LB	Pound
LVB	Large Vehicle Bomb
MK	Mark
MNSA	Maritime Non-State Actor

MM	Millimeter
MOD	Ministry of Defence
MOUT	Military Operations in Urban Terrain
MPF	Ministerio Público Federal; Federal Public Ministry
MPH	Miles Per Hour
MS-13	Mara Salvatrucha 13
MXN	Peso Currency Code
NFZ	No Fly Zone
NIAC	Non-International Armed Conflict
OE	Operational Environment
OOB	Order-Of-Battle
OZ	Ounce
PEP	Policía Estatal Preventiva; State Preventive Police
PF	Policía Federal; Federal Police
PMSC	Private Military and Security Company
PSYOPS	Psychological Operations
RF	Radio Frequency
RPG	Rocket Propelled Grenade
SASO	Stability and Support Operations
SEC	Second
SEDENA	Secretaría de la Defensa Nacional; Secretariat of National Defense
SEIDO	Subprocuraduría Especializada en Investigación de Delincuencia Organizada; Specialized Office of the Special Prosecutor for Organized Crime Investigation
SEMAR	Secretaría de Marina; Secretariat of the Navy
SIM	Subscriber Identification Module

SSP	Secretaría de Seguridad
SSPE	Secretario de Seguridad Pública Estalal; Public Safety Secretary
sUAS	Small Unmanned Aircraft System
SUAS	Single Unmanned Aircraft System
sUAV	Small Unmanned Aerial Vehicle
SUV	Sport Utility Vehicle
TCO	Transnational Criminal Organization
TNT	Trinitrotoluene
TTPs	Tactics, Techniques, & Procedures
TV	Television
UAS	Unmanned Aerial System
UAV	Unmanned Aerial Vehicle
UGV	Unmanned Ground Vehicle
UMV	Unmanned Marine Vehicle
US	United States
USD	US Dollar
VBIED	Vehicle Borne Improvised Explosive Device
VNSA	Violent Non-State Actor
VR	Virtual Reality
WWII	World War II

Preface

Why Drones – Why Now?

David Hambling

London, England

August 2021

The modern consumer drone industry took off in 2014. Perhaps the chief architect of the quadcopter revolution is Frank Wang, an entrepreneur who founded Da-Jiang Innovations—better known as DJI—in Shenzhen, China. Wang's goal was simply to provide a method for low-cost aerial photography that anyone could use.

Radio-controlled aircraft have been flown as a hobby since the 1940s. Hobbyists enjoy aircraft for the sake of aircraft. Their interest is in building, displaying and flying their models. Construction is often laborious, with painstaking attention to detail over scale models of historical aircraft.

Learning to fly a radio-controlled model requires the same type of understanding as flying a light aircraft. Becoming a skilled remote-control pilot takes many hours of practice and, usually, a large number of crashes. It is challenging, which is a great part of the appeal.

By contrast, drones are simply platforms for airborne cameras. DJI chose the quadcopter for its simplicity, without the need for a tail rotor or adjustable main rotor of a helicopter. Steering, forward flight and hovering can be accomplished by adjusting the speed of the four rotors, all of which are handled by the electronics. The small blades of the quadcopter are safer and less likely to strike objects, making it more suitable for cluttered urban airspace or even indoor flying.

Wang quickly realized that his company had to assemble all of the aerial system elements, many of which did not exist at that time. Drones were starting to appear, but they needed a cheap, effective flight controller—the electronics that run the autopilot so all the operator has to do is point the drone in the right direction. Another missing part was a lightweight gimbal, the swivel mounting that keeps the camera pointed in the same direction and compensates for the drone's motion.

Wang's team created these and more. DJI are tightly integrated with suppliers: rather than using off-the-shelf components, all the electronics are customized to their exact requirements and optimized for size, weight, and power. The end product was a remote-control craft which could be flown, out of the box, with no skill at all. To make it hover in place, you just press a button; to bring it back to make a safe landing, press another button. Boring for flying enthusiasts, but exciting for photographers and film-makers.

Drone photography was revolutionary. Previously, the only way of getting aerial shots was with the aid of a crane or a helicopter, both of which are expensive pieces of kit confined very much to the professional end of the field. Now, everyone had access to a stable aerial platform that could fly anywhere.

The original drones were sized to carry a miniature camera weighing tens of grams. The cheapest, like the DJI Mavic, cost just a few hundred dollars. Some users wanted to carry something bigger, like the video cameras used in the movie and TV business and the market soon catered for them with craft like the DJI Inspire 2.

Now low-budget indie directors could employ swooping visuals over a landscape before zooming in on a character standing on a balcony or could fly over a car chase. Every documentary or TV advert could include drone footage and drone photography soon became a common extra for wedding videos and real estate websites.

The drone market took off faster than anyone had anticipated. In 2013, the FAA (Federal Aviation Administration) estimated that there might be "as many as 30,000" civilian drones operating in the US in 2020. By 2018, there were an estimated 1.25 million.

DJI has dominated the drone market from the outset, currently accounting for some 76% of drone sales in the US. The second place is held by Intel, with an estimated 4%. Their previous US competitors, 3D Robotics, left the drone business some years ago.

DJI succeeded beyond expectation in providing drones for everyone, and everyone includes nefarious actors. Being able to fly over walls, fences, and other barriers gave access to areas which were impossible to reach by other means. Before long, there were stories about drone paparazzi intruding on celebrities in their homes, drones smuggling contraband into prisons, and drones being modified to carry weapons.

The drones involved were typically off-the-shelf models with minimal modifications, for example simply duct-taping contraband to the underside. Others were making more elaborate modifications, such as adding a simple drop mechanism to release a payload. And, at the more advanced end, there were some who were building their own drones, usually fixed-wing types resembling small aircraft, for illicit purposes.

While, in theory, it would have been possible for any criminal to master the construction and piloting of radio-controlled aircraft in the 1970s, 80s or 90s, this never happened. The dedication, not to mention the number of crashes involved, were an effective deterrent to misuse. In addition, radio-controlled aircraft were rare and in the hands of a small community who all tended to know each other.

In the early years of drone adoption, the most conspicuous repurposers of commercial drones were the Islamic militant group ISIS (Islamic State of Iraq and al Sham) in Iraq and Syria. The group had always been known for slickly-produced propaganda videos, and when everyone else was adding impressive aerial shots to their videos, ISIS followed suit, shooting VBIEDs (Vehicle Borne Improvised Explosive Devices) rather than weddings. In the next few years, they graduated swiftly through using drones for reconnaissance to directing mortar fire and guiding VBIEDs through urban areas. Then there were the first reports of drones booby-trapped with explosives and, finally, grenade-dropping drones.

ISIS had a natural advantage in armed drone development as they had access to a full range of military hardware. Consumer drones have a limited payload capacity, but a 40mm grenade, which weighs about two hundred grams, turned out to be a good match. Soon ISIS was putting up videos from drones dropping grenades on a variety of targets with considerable accuracy.

The use of such modified drones spread rapidly across the region, not just among other insurgent groups but also by the Iraqi Federal Police. The US military, although well-supplied with its own drones, also showed some interest, especially for training against 'aggressor forces.' However, concerns about the security of using Chinese technology later led to a Pentagon-wide ban on DJI and other Chinese-made drones.

Given this global background of drone use and repurposing for nefarious purposes, it is unsurprising that Mexican cartels adopted them, but not quite inevitable. The potential advantages of drones must be offset against the challenges of introducing new and untried technology, especially where it has a potential downside. Just as the US military was concerned about Chinese government collecting intelligence data from DJI drones, cartels may have similar doubts over whether drones might be feeding data to the authorities.

As this book edited by Dr. Robert J. Bunker and Dr. John P. Sullivan makes clear, Mexican cartel use of drones is now well-established and, although the spread has been slow, it has been steady. This is not a technology fad which will go away. Drones are now part of the challenge faced by law-enforcement agencies. As with counter-insurgency elsewhere in the world, they are a challenge that is only likely to become more serious in the coming years as DJI and other produce ever more sophisticated drones with increasing levels of onboard intelligence.

Foreword

Weaponized Drones and Order-Of-Battle (OOB) Criteria

Lisa J. Campbell

Ventura, California

August 2021

On 18 April 2021, Iran held a military parade marking its National Army Day. Included in this showy display of Iranian military firepower was a large selection of drones of varying sizes and capabilities. The rapid emergence and ubiquitous presence of drones within Iran's military is indicative of the rising appeal of the capability to a wide variety of actors worldwide, including within the illicit world of Mexican cartels. The use of drones by cartels in Mexico has evolved since roughly 2010, beginning merely as a means to deliver drugs across the border. From there, usage expanded to include capabilities of intelligence, surveillance and reconnaissance (ISR) and more recently, weaponization. Dr. Robert Bunker and Dr. John Sullivan are at the forefront of this evolving threat, tracking and assessing the activity as it continues to develop.

Drones have proven to have significant appeal to non-state actors because of their multi-functional uses that satisfy many order-of-battle (OOB) criteria. Some operational examples of OOB uses and/

or advantages with drones include command and control, tactics, technology, equipment, resupply, logistics, personnel numbers (i.e., the drone as a robot replacing a person), electronic combat, etc. This anthology discusses emerging drone tactics unique to the cartels as well as those borrowed from users such as the Islamic State in the Middle East. Mexican cartels, for example, have used swarming tactics employing cars and trucks with the purpose of conducting assassinations or theft of commercial goods. This tactic could be adapted with drones to meet similar or new objectives. Drone systems are easily transitioned by adversarial militaries from conventional use to function asymmetrically,[1] and there are recent examples around the world where experimental use of drones has resulted in one-sided tactical advantages in warfare.

In late 2020, for example, the most recent battle of the decades-long conflict between Armenia and Azerbaijan demonstrated the surprising effect drones can have on an unprepared adversary. During the six-week long conflict, Azerbaijan made territorial gains in the disputed Nagorno-Karabakh territory, which subsequently reshaped the dynamic there.[2] Aided by new revenue streams and support from Turkey, Azerbaijan fielded several types of drones, including highly effective attack systems obtained from Turkey and Israel. By contrast, Armenia's drone fleet consisted of smaller indigenous systems used almost exclusively for conducting reconnaissance missions. Armenia's air defenses in Nagorno-Karabakh were older Soviet or Russian systems that were largely ineffective against Azerbaijan's drone tactics.[3] It is widely assessed that the effective use of weaponized drones was instrumental in turning the tide of the war in favor of Azerbaijan. Azerbaijan had, in effect, leveraged known tactics, techniques, and procedures (TTPs) using an unmanned fleet, to destroy Armenian defenses and establish tactical air superiority with minimal risk.[4] The outcome of this latest battle got the attention of global powers such as Russia and Turkey, who are reportedly examining the lessons learned.

At least two lessons learned observed by analysts of the 2020 Armenia-Azerbaijan conflict are noteworthy: "Without adequate sensors, electronic warfare cover, and counter-drone weaponry, traditional ground units are in trouble"[5] and "Integration of land-based fire support and drones looms large in modern warfare."[6] The recent Azerbaijani-Armenian clashes revealed the vulnerability of traditional land units in the face of advanced drone warfare weaponry and concepts. Integration of land-based fire with drones has also been recently demonstrated in Syria.

A similar drone use evolution is emerging in Mexico like those in other parts of the world, and it warrants close analytical scrutiny. There are indications that cartels, particularly *Cártel de Jalisco Nueva Generación* (CJNG), are seeing the utility of integrating drones with other land-based fire, such as armed gunmen or up-armored vehicles with shooters. More importantly, however, is the indication that, without drones or drone counter-measures, the rivals of CJNG are likely going to be outmatched. CNJG's use of drones currently outweighs that of other peer illicit groups and even the various Mexican security services. While still in the early stages of weaponizing drones, the combat advantages for CJNG are becoming apparent, and this should drive its adversaries to seek drone-related combat options of their own. With an effective weaponized drone arsenal, CJNG will likely meet its objectives of control of territory, perpetual illicit business, controlling the population, distracting or disrupting government security services, etc. until adequately opposed. Any opposing force, whether it be rival cartels, Mexican military or police, likely has two options: acquire a comparable drone fleet of their own or incorporate drone countermeasures into their arsenals.

Dr. Bunker, Dr. Sullivan, and additional researchers accurately predicted that cartels would eventually up-gun drones to use as weapons against their adversaries, well before the first evidence of their use was found in Mexico. Tracking this evolution via "Tactical Notes" and other assessments throughout this anthology, the authors provide a sense that drone use by cartels has only just begun while there remains unlimited upside potential, given commercial availability and fast- growing technology. Dr. Bunker and Dr. Sullivan in this new anthology assess that the use of drones for states and their competitors is no longer merely a tactical issue, but has strategic and operational potentials. These potentials apply to a growing number of geographical areas and levels of conflict, including the US-Mexico border region. Along the border, a segment of airspace previously little used is becoming contested with a large presence of drones, including those employed by US and Mexican agencies. Drones in such areas can enable illicit actors to gain air superiority in unconventional ways. For example, in 2011, US authorities shot down multiple drones entering the US illegally from Mexico, but overall, the cartels were unimpeded due to the vast border area, ability to avoid some radar detection, and by having a virtually unlimited re-supply of drones; thus, the cartels were able to confound and defeat the US effort.

The potential use of technology discussed in this book is emerging already, is available globally, at low cost, and with relatively low expertise required. Criminal organizations who operate in war-like zones will not have to rely on obtaining military-grade drones when the civilian technology is equal if not better. A key outcome of tracking adversary operational capabilities is to be able to detect 'red flags,' that is, causal alerts when changes in an adversary's OOB has potential to impact that of friendly forces. This anthology provides the capability to detect such red flags as they relate to drones in Mexico and at the southern US border; it is a thorough operational study that includes realistic estimates of future uses of drones by Mexican cartels. Cartel drones are complicating the US-Mexican border area already with narcotics smuggling and ISR; eventually lethal drones may inhabit the border area as well. The US and Mexican governments may eventually be driven to decision-points on where and when to effectively place resources while prioritizing which illicit drone operation to counter: drug smuggling, ISR and/or weaponized. As cartels evolve this capability, and those decisions become more critical, this anthology may aid in the efforts to provide a more accurate picture of these developing threats.

Endnotes

[1] Iran and its proxy forces in the Middle East—being outmatched by the US militarily—utilize drones in unconventional ways against US and allied nations in an attempt to drive the US and coalition out of the region.

[2] Congressional Research Service, "Azerbaijan and Armenia: The Nagorno-Karabakh Conflict." 7 January 2021, Azerbaijan and Armenia: The Nagorno-Karabakh Conflict (congress.gov).

[3] Ibid.

[4] Nicole Thomas, LTC Matt Jamison, CAPT(P) Kendall Gomber, and Derek Walton, "What the United States Military Can Learn from the Nagorno-Karabakh War." *Small Wars Journal*. 4 April 2021, https://smallwarsjournal.com/index-php/jrnl/art/what-united-states-military-can-learn-nagorno-karabakh-war.

[5] "Five key military takeaways from Azerbaijani-Armenian war: analysis." *YeniSafak News*. 30 October 2020, https://www.yenisafak.com/en/news/five-key-military-takeaways-from-azerbaijani-armenian-war-analysis-3552970.

[6] Ibid.

Introduction

Criminal Armed Groups and Drones – from Smuggling and ISR, to Improvised Weapons Platforms

John P. Sullivan and Robert J. Bunker

Los Angeles, California

September 2021

Drones have become part of the contemporary landscape of conflict and crime. Like all technological developments, they start small and limited and then spread as their advantage becomes clear to their expanding user base. In this text, we explore the proliferation of aerial drones—or Unmanned Aerial Systems (UAS)—among the Mexican Cartels and gangs.[1] That is, we document and assess the evolution of aerial drones by these non-state criminal actors as methods and means to further their goals. Since Mexican cartels are involved in a range of criminal enterprises and cross-border activity at the US-Mexico border, we see early cartel drone use focused on smuggling drugs. After all, the cartels are—at least in part—drug trafficking organizations (DTOs). Beyond being used as modes for smuggling contraband and narcotics across the border, drones have been used to enhance situational awareness and are employed as aerial intelligence, surveillance, and reconnaissance (ISR) platforms to allow the criminal enterprises to thwart law enforcement and border security counter measures. Of course, border security agencies (i.e., Border Patrol, Customs and Border Protection), along with their law enforcement and military counterparts on both sides of the frontier, also use UAS to monitor the border. Finally, these criminal enterprises—which battle each other and the state for freedom of action—have been adapting commercial-off-the-shelf (COTS) drones or small UAS (sUAS) as weapons systems in pursuit of combat supremacy in their interlocked criminal insurgencies and crime wars.

Illicit Tactical Progress—Drones

The ability to integrate new technologies into competitive enterprises like conflict (war), commerce, and crime confers advantage. Indeed, technological innovation has long been seen as a desirable military trait. Technological superiority is sought after to ensure strategic success. Classical technological developments include the use of walls, fortifications, and engineering to protect the populace (and control bandits), gunpowder and firearms (including cannons), and explosive devices. The side that best integrated these new technologies (like the archers at Agincourt) often prevail.[2] In the realm of terrorism, bombs and explosives became a staple after Alfred Noble developed dynamite. This allowed the anarchists and revolutionaries of the 19th Century greater combat power and lethality facilitating the "Propaganda of the Deed."[3][4] Mario Buda, ushered in the era of car bombs (including large vehicle bombs or LVBs and vehicle-borne improvised explosive devices or VBIEDS) when he blew up a wagon packed with explosives at the corner of Wall and Broad Streets in New York's Financial District in an anarchist bombing in 1920.[5] Car bombs and other explosives (Improvised Explosive Devices or IEDs) have dominated terrorism and guerrilla warfare ever since. Indeed, they even punctuate Mexico's criminal insurgencies.[6]

Technological and tactical progress is clearly evident in the increased sophistication of criminal actors in Mexico's crime wars.[7][8] Here, the violent conflict extends beyond typical criminal violence and *sicarios* (hitmen) use, to embrace terrorist tactics, infantry operations, and barbarization in a quest for cartel territorial control, power, and profit.[9] At times, the levels of sustained violence and sophistication of the groups involved reaches levels that rightfully become non-international armed conflicts (NIACs); acknowledged or not.[10]

Increasingly, aerial drones are becoming embedded components of contemporary irregular conflict.[11] Terrorist, insurgents, guerrillas, and criminals (criminal armed groups or CAGs) find the flexibility of aerial drones appealing. Homemade or artisanal drones now threaten conventional armed forces and police. Guerrillas such as the Islamic State (IS) have used drones against Russian forces in Syria since 2015.[12] Use of drones by violent non-state actors (VNSAs) is on the rise and largely underappreciated, with Hamas, Hezbollah, and Houthi rebels joining IS in their use.[13] Aerial drones have been used as assassination weapons as seen in the 2018 attempt on Venezuela's president Nicolás Maduro[14] and, more recently, the assassination of Haiti's president on 7 July 2021.[15] In June 2021, two explosive-laden drones attacked an Indian Air Force base in Jammu on 26 June 2021. A day later, the Indian Army spotted two drones loitering near its Kaluchak station overnight on 27-28 June 2021.[16] On the criminal side of the house, Brazilian gangsters used drones, along with explosives and human shields, in a 'New *Cangaço*' style urban bank raid near São Paulo[17] and drug cartels, as recently as 13 September 2021 when three drones dropped their explosive payload on the prison in Guayaquil, have attacked Ecuadorian prisons targeting rival gang leaders as part of their inter-cartel rivalry.[18]

Charting the Course of Criminal Drone Evolution

This text reviews cartel drone evolution in Mexico and Latin America (with discussions of drone assaults and use in Colombia and

Venezuela augmenting the Mexican cases). We believe the case studies described here illustrate the technological and tactical interface that fuels combat innovation by criminal organizations. As such, this text complements our recent book *Illicit Tactical Progress: Mexican Cartel Notes 2013-2020* to document the development of cartel tactics, arms, and irregular warfighting capabilities.[19]

Criminal Drone Evolution: Cartel Weaponization of Aerial IEDs starts charting the course of criminal use of aerial drones with a preface by journalist David Hambling, author of *Weapons Grade* (2005), *Swarm Troopers: How Small Drones Will Conquer the World* (2015), and *We: Robot* (2018), to set the stage for examining the disruptive potentials of small unmanned aircraft. The text continues with an assessment of drones within criminal orders-of-battle (OOB) by US military intelligence officer Lisa J. Campbell followed by this introduction to close out the front matter.

The main text contains 22 chapters documenting the evolution of drone use in Mexico's competitive narco-conflict ecology. Chapter 1 by Camilo Mejia Giraldo (2014) is on the DEA's discovery of custom narco-drones. It is followed by Chapter 2: Mexican Cartel Tactical Note #21 by Robert J. Bunker, examining Cartel UAVs (2014), and Chapter 3 an *OE Watch* note by Brenda Fiegel, looking at drones used for cross-border smuggling (2014). Chapter 4 from U.S. Customs and Border Protection is a *CBP Newsroom* (2016) release that documents marijuana smuggling by drone in November 2015 Arizona.

Chapter 5 by John P. Sullivan and Robert J. Bunker, an analysis of criminal drone potentials (2016), looks at narco-drones on the border and beyond. A short *CBP Newsroom* release (2016) provides case data on drone activity in the Border Patrol's Yuma, Arizona sector in Chapter 6. This is followed by a short analysis (2017) from *Small Wars Journal* by Brenda Fiegel on drones and drug trafficking (Chapter 7). Her analysis is followed by Chapter 8 in which another CBP case report (2017) on drone smuggling near San Diego's San Ysidro Port of Entry is discussed.

In Chapter 9, Robert J. Bunker and John P. Sullivan, document weaponized drone deployment in Guanajuato (2017) where remote detonation IEDs using *'papa bomba'* payloads were interdicted by

Mexican police. Documentation of this drone weaponization trend continues in Chapter 10 where Sullivan and Bunker are joined by David A. Kuhn to dissect the targeting of the Baja California Public Safety Secretary's residence in Tecate, BC (2018). Parker Asmann at *InSight Crime* provides a synopsis of this incident in Chapter 11.

The use of drones for ISR (recce and route recon) is documented in Chapter 12 from *CBP Newsroom* (2019) while Chapter 13 from *Borderland Beat* captures the use of drones to record humanitarian aid distribution by the *Cártel Jalisco Nueva Generación* (CJNG). Drone smuggling in the Border Patrol's Yuma Sector is again reported in Chapter 14 by *CBP Newsroom* (2020). Chapter 15 returns to a weaponized drone focus when Kuhn, Bunker, and Sullivan recount the seizure of drones and explosives in Puebla by the *Fiscalía General de la República* (FGR) and *Secretaría de la Defensa Nacional* (SEDENA) related to the contest between the *Cártel de Santa Rosa de Lima* (CSRL) and its rivals in Guanajuato. Chapter 16 from the *Policía Comunitaria Tepalcatepec* recounts the presence of 'El Chino Drones,' a former marine (sailor or member of the naval infantry) as a bombmaker for the CJNG.

Chapter 17 from Katie Jones at *InSight Crime* (2020) provides an overview of organized crime exploitation of drones to further their goals with cases discussing weaponized drones use by the CJNG in Tepalcatepec, Michoacán; illicit loggers, drug traffickers, and gangs using drones for ISR in Brazil and El Salvador (MS-13); and traffickers using drones for close-in air traffic control in Guatemala. The CJNG incident in Tepalcatepec is assessed in Chapter 18 (2020) by Bunker, Sullivan, Kuhn, and Alma Keshavarz. Chapter 19 (2020) once again recaps drone smuggling in the Yuma, Arizona, Border Patrol sector.

The use of weaponized drones/aerial-IEDs culminates in the following chapters, with Chapter 20 by Bunker and Sullivan recounting the use of aerial drones for ISR in a CJNG engagement of *autodefensa* (self-defense force) mounted infantry in improvised armored fighting vehicles (IAFV) in La Bocanda Michoacán (2020) and Chapter 21, by Bunker and Sullivan that recounts an alleged CJNG drone attack against two police officers in Aguililla, Michoacán (2021). Chapter

22, again by Bunker and Sullivan, (2012) documents additional cartel drone activity, including a CJNG attack and the arrest of CSRL drone weaponeers linked to an earlier incident.

Conclusions and Future Potentials

Criminal Drone Evolution concludes with a summation of the text and its findings by Bunker and Sullivan. The conclusion focuses on Mexican cartel—specifically *Cártel Jalisco Nueva Generación* (CJNG)—drone utilization and combat trends related to multi-use weaponized drones, propaganda videos, ISR integrated into combined arms operations, and night vision capability. It is then augmented with an Afterword by red teaming and drone specialists Conrad 'Andy' Dreby and Scott Crino on UAS potentials and a Postscript by James J. Torrance, a US Army Signal Corps officer, on anticipating future unmanned systems threats. These threats include moving beyond aerial drones to biomimicry, hacking self-driving cars, and 3D printing. These insights are complemented by five appendices: Appendix 1 by Brenda Fiegel, a senior intelligence analyst at the Foreign Military Studies Office (FMSO) at Ft. Leavenworth looks at narco-drones in Colombia (2017), while Bunker and Keshavarz (2018) examine the alleged assassination attempt on Venezuelan president Maduro in Appendix 2. Maria Alejandar Navarrette, an *InSight Crime* analyst, resumes the assessment of Colombia, this time on its Pacific Coast (2019) in Appendix 3. Appendix 4 by *La Seguridad es de Todos* looks at drone proliferation in Nariño, Colombia (2019). Brenda Fiegel returns with a short note (2020) in Appendix 5 recapping the experience of drug traffickers in Brazil, Colombia, and Peru while adding comments on Mexican cartel drone use.

Together, this collection documents the evolution of drone usage by criminal cartels, gangs, and criminal armed groups (CAGs) in Mexico and beyond. It demonstrates the development spread of drone technology and tactics, techniques, and procedures (TTPs) among criminal actors while also pointing toward future potentials.[21]

Endnotes

[1] See Robert J. Bunker, John P. Sullivan, and David A. Kuhn. "Use of Weaponized Consumer Drones in Mexican Crime War." *Counter-IED Report*. Winter 2020-21, pp. 69-77, available at https://www.academia.edu/45000611/Use_of_Weaponized_Consumer_Drones_in_Mexican_Crime_War.

[2] The proficient use of longbows allowed the Henry V to overcome a numerically superior French force at the Battle of Agincourt, 25 October 1415. The battle is chronicled by William Shakespeare in his play *Henry V.*

[3] See Simon Werrett, "The Science of Destruction: Terrorism and Technology in the Nineteenth Century" in Carolina Dietze and Claudia Vernhoeven, Eds., *The Oxford Handbook of the History of Terrorism*, Oxford: Oxford University Press. January 2014: online, https://www.oxfordhandbooks.com/view/10.1093/oxfordhb/9780199858569.001.0001/oxfordhb-9780199858569-e-012.

[4] See Constance Bateman, "The Era of the Propaganda of the Deed" in Carl Levy and Matthew S. Adams, Eds., *The Palgrave Handbook of Anarchism*. London: Palgrave Macmillan Cham. 2018: pp. 371-387, https://doi.org/10.1007/978-3-319-75620-2_22.

[5] Mike Davis, *Buda's Wagon: A Brief History of the Car Bomb*. New York: Verso, 2007.

[6] See Robert J. Bunker and John P. Sullivan, "Cartel Car Bombings in Mexico." *The Letort Papers*. Carlisle Barracks: Strategic Studies Institute, US Army War College, August 2013, https://publications.armywarcollege.edu/pubs/2238.pdf.

[7] On criminal insurgency, see for example, John P. Sullivan, "From Drug Wars to Criminal Insurgency: Mexican Cartels, Criminal Enclaves and Criminal Insurgency in Mexico and Central America. Implications for Global Security." *Working Paper Nº 9*, April 2012. Paris: Fondation Maison des sciences de l'homme, https://halshs.archives-ouvertes.fr/halshs-00694083/document.

[8] On cartel tactical progress, see John P. Sullivan, "Mexican Cartel Adaptation and Innovation." *OODA Loop*. 27 January 2020, available at https://www.academia.edu/41754168/Mexican_Cartel_Adaptation_and_Innovation.

[9] See David Teiner, "Cartel-Related Violence in Mexico as Narco-Terrorism or Criminal Insurgency: A Literature Review." *Perspectives on Terrorism*. Vol. 14, no. 4. August 2020: pp. 83-98, available at https://www.jstor.org/stable/26927665.

[10] See "Non-International Armed Conflicts in Mexico." *RULAC: Rule of Law in Armed Conflicts*. Geneva: Geneva Academy of International Humanitarian law and Human Rights. Multiple dates, updated 30 May 2021, https://www.rulac.org/browse/conflicts/non-international-armed-conflict-in-mexico.

[11] See Robert J. Bunker, *Terrorist and Insurgent Unmanned Aerial Vehicles: Use, Potentials, and Military Implications*. Carlisle Barracks: Strategic Studies Institute, US Army War College, 2015, https://publications.armywarcollege.edu/pubs/2238.pdf.

[12] "Home-made drones now threaten conventional armed forces." *The Economist.* 8 February 2018, https://www.economist.com/science-and-technology/2018/02/08/home-made-drones-now-threaten-conventional-armed-forces.

[13] See Kerry Chávez and Ori Swed, "Off the Shelf: The Violent Nonstate Actor Drone Threat." *Air & Space Power Journal.* Fall 2020, https://www.airuniversity.af.edu/Portals/10/ASPJ/journals/Volume-34_Issue-3/F-Chavez_Swed.pdf.

[14] Nicholas Grossman, "Are drones the new terrorist weapon? Someone tried to kill Venezuela's president with one." *Washington Post.* 10 August 2018, https://www.washingtonpost.com/news/monkey-cage/wp/2018/08/10/are-drones-the-new-terrorist-weapon-someone-just-tried-to-kill-venezuelas-president-with-a-drone/.

[15] Jacqueline Charles and Jay Weaver, "Grenade-dropping drones, a paranoid president, guards who ran: Latest on Haiti assassination." *Miami Herald.* 19 September 2021, https://www.miamiherald.com/news/nation-world/world/americas/article254275213.html.

[16] See Neeraj Chauhan, "LeT planned successive drone attacks in J&K: Intel." *Hindustan Times.* 27 August 2021, https://www.hindustantimes.com/india-news/let-planned-successive-drone-attacks-in-j-k-intel-101630001006088.html; Alijaz Hussain, "Drone Attacks on Indian Air Force Base in Jammu Underscore New Threat." *The Diplomat.* 28 June 2021, https://thediplomat.com/2021/06/drone-attacks-on-indian-air-force-base-in-jammu-underscore-new-threat/; and Kamaljit Kaur Sandu, "Day after attack on Jammu air base, Army spots 2 drones over Kaluchak military station | 10 points." *India Today.* 28 June 2021, https://www.indiatoday.in/india/jammu-and-kashmir/story/jammu-drone-attack-air-force-station-ied-probe-update-development-1820181-2021-06-28.

[17] John P. Sullivan, José de Arimatéia da Cruz, and Robert J. Bunker, "Third Generation Gangs Strategic Note No. 42: Brazilian Gangs Utilize Human Shields, Explosives, and Drones in a New 'Cangaço' Style Urban Bank Raid in Araçatuba, São Paulo." *Small Wars Journal.* 5 September 2021, https://smallwarsjournal.com/jrnl/art/third-generation-gangs-strategic-note-no-42-brazilian-gangs-utilize-human-shields.

[18] See Bruce Crumbly, "Drones drop explosives in Ecuador prison attack by suspected drug cartels." *DroneDJ.* 14 September 2021, https://dronedj.com/2021/09/14/drones-drop-explosives-in-ecuador-prison-attack-by-suspected-drug-cartels/.

[19] Robert J. Bunker and John P. Sullivan, Eds. *Illicit Tactical Progress: Mexican Cartel Tactical Notes 2013-2020.* (A Small Wars Journal-El Centro Anthology.) Bloomington: Xlibris, 2021 and an interview of Drs. Bunker and Sullivan on cartel tactical evolution in Chris Dalby, "How Mexico's Cartels Have Learned Military Tactics." *InSight Crime.* 2 September 2021, https://insightcrime.org/news/how-mexicos-cartel-have-learned-military-tactics/.

[20] For a parallel discussion of the dissemination of cartel tactical innovation (TTPs), see Nathan P. Jones, "Bacterial Conjugation as a Framework for the

Homogenization of Tactics in Mexican Organized Crime." *Studies in Conflict & Terrorism*. Vol. 44, no. 10: pp. 855-884, https://doi.org/10.1080/10576 10X.2019.1586356.

[21] While future potentials are discussed in depth in the conclusion to this work, it is worth noting that these potentials include continued technical sophistication, including the integration of artificial intelligence (AI) and killer robot swarms to enable VNSAs and CAGs to field lethal autonomous weapons (LAWS). This may presently seem extreme, but the proliferation of lethal capabilities always seems far off until it is employed in actual combat. See Jacob Ware, "Terrorist Groups, Artificial Intelligence, and Killer Drones. *War on the Rocks*. 24 September 2019, https://warontherocks.com/2019/09/terrorist-groups-artificial-intelligence-and-killer-drones/; Zachary Kallenborn. "The Era of the Drone Swarm is Coming, and We need to be Ready for It." *Modern War Institute*. 25 October 2018, https://mwi.usma.edu/era-drone-swarm-coming-need-ready/; and Zachary Kallenborn and Phillip C. Bleek, "Drones of Mass Destruction: Drone Swarms and the Future of Nuclear, Chemical, Biological Weapons. *War on the Rocks*. 14 February 2019, https://warontherocks.com/2019/02/drones-of-mass-destruction-drone-swarms-and-the-future-of-nuclear-chemical-and-biological-weapons/.

Chapter 1

Mexico's Cartels Building Custom-Made Narco Drones: DEA

Camilo Mejia Giraldo

Initially Published in InSight Crime on 11 July 2014

Mexico's drug cartels are reportedly commissioning custom-made drones to transport narcotics across the US border, illustrating the continual development of innovative new technologies and methods used to traffic drugs.

Cartels have begun hiring local workers from companies in Mexico to develop custom drones, or unmanned aerial vehicles (UAV), suited to their needs, according to an unnamed Drug Enforcement Administration (DEA) source consulted by El Universal.[1]

Since 2012, the DEA has registered around 150 narco drones crossing the border, transporting in total approximately two tons of cocaine and other drugs. This amounts to an average of roughly 13 kilos per load.

US and Mexican authorities have identified Queretaro, Guadalajara, Nuevo Leon and Mexico City as the drone production points, where cartels pay professionals two to three times their normal salary for this custom technology.

While cartels used to use foreign-made drones, the new reliance on home-grown technology and construction is more cost-effective—the method is cheaper than the construction of cross-border tunnels or the use of semi-submersible vessels, according to El Universal's report.

InSight Crime Analysis

These new developments represent the latest in a long list of technologies and drug transport methods developed by cartels.

Clandestine tunnels, which drug trafficking organizations have used since at least 1990,[2] have become increasingly sophisticated[3] in recent years. A "super tunnel" discovered in 2013, for example, used a railway line to transport drugs and was equipped with electricity and ventilation.

The transformation of the semi-submersible craft is also indicative of this trend.[4] Initially rudimentary in their design to facilitate trafficking in maritime routes, they have been developed to sport kitchens and air-conditioning, with the capacity to carry around eight tons of cocaine.

In addition to drones, cartels use other forms of aerial transport, like ultralight aircraft,[5] an inexpensive method favored by drug trafficking organizations for their ability to evade radar while carrying around 100 kilo loads.

Mexican criminal organizations have been using UAVs since at least 2010,[6] but the relatively small amount of drugs transported per trip on the drones registered by the DEA helps explain why the cartels want to develop larger, specially tailored UAVs. As drone technology worldwide becomes increasingly accessible and cost effective, commissioning custom-made drones in Mexico is the next logical step for cartels.

Endnotes

[1] Doris Gómora, "Fabrican narcos sus propios drones, altera la DEA." *El Universal.* 9 July 2014, https://archivo.eluniversal.com.mx/nacion-mexico/2014/carteles-fabrican-narcodrones-trasiego-eu-1022274.html.

[2] Martha Maguire, "US Warns of Rise in Secret Tunnels Under Mexico Border." *InSight Crime.* 16 Jun 2011, https://www.insightcrime.org/news/brief/us-warns-of-rise-in-secret-tunnels-under-mexico-border/.

[3] Miriam Wells, "US Shuts Down Mexico Narco 'Super Tunnel.'" *InSight Crime*. 1 November 2013, https://www.insightcrime.org/news/brief/us-shuts-down-narco-super-tunnel-mexico/.

[4] Hannah Stone, "RawFeed: The Evolution of the Drug Submarine." *InSight Crime*. 8 March 2011, https://www.insightcrime.org/news/analysis/the-evolution-of-the-drug-submarine/.

[5] Ronan Graham, "US Cracks Down on Drug Smugglers in Ultralight Planes." *InSight Crime*. 15 December 2011, https://www.insightcrime.org/news/brief/us-cracks-down-on-drug-smugglers-in-ultralight-planes/.

[6] "'Narco drones' puts all U.S. border efforts in question." *The American Post*. 2010, https://www.theamericaspostes.com/2338/narco-drones-puts-all-u-s-border-efforts-in-question/.

Chapter 2

Mexican Cartel Tactical Note #21: Cartel Unmanned Aerial Vehicles (UAVs)

Robert J. Bunker

Initially Published in Small Wars Journal on 1 August 2014

Key Information: Camilo Mejia Giraldo, "Mexico's Cartels Building Custom-Made Narco Drones: DEA." *Insight Crime.* 11 July 2014, http://www.insightcrime.org/news-briefs/mexicos-cartels-building-custom-made-narco-drones-dea:

> Mexico's drug cartels are reportedly commissioning custom-made drones to transport narcotics across the US border, illustrating the continual development of innovative new technologies and methods used to traffic drugs.
>
> Cartels have begun hiring local workers from companies in Mexico to develop custom drones, or unmanned aerial vehicles (UAV), suited to their needs, according to an unnamed Drug Enforcement Administration (DEA) source consulted by El Universal.

Since 2012, the DEA has registered around 150 narco drones crossing the border, transporting in total approximately two tons of cocaine and other drugs. This amounts to an average of roughly 13 kilos per load.

US and Mexican authorities have identified Queretaro, Guadalajara, Nuevo Leon and Mexico City as the drone production points, where cartels pay professionals two to three times their normal salary for this custom technology.

While cartels used to use foreign-made drones, the new reliance on home-grown technology and construction is more cost-effective – the method is cheaper than the construction of cross-border tunnels or the use of semi-submersible vessels, according to El Universal's report.

InSight Crime Analysis

These new developments represent the latest in a long list of technologies and drug transport methods developed by cartels...

Mexican criminal organizations have been using UAVs since at least 2010, but the relatively small amount of drugs transported per trip on the drones registered by the DEA helps explain why the cartels want to develop larger, specially tailored UAVs. As drone technology worldwide becomes increasingly accessible and cost effective, commissioning custom-made drones in Mexico is the next logical step for cartels.

Key Information: Doris Gómora, "Fabrican narcos sus propios drones, alerta la DEA." *El Universal.* 9 de Julio de 2014, http://www. eluniversal.com.mx/nacion-mexico/2014/carteles-fabrican-narcodrones-trasiego-eu-1022274.html:

Cárteles del narcotráfico están fabricando en
México sus propias aeronaves no tripuladas o drones
para traficar droga hacia los Estados Unidos, para lo
cual están utilizando a trabajadores de empresas que
tienen ensambladoras de drones en territorio mexicano,
según información de la Agencia Antidrogas de Estados
Unidos (DEA).

Narcodrones han sido utilizados para traficar droga
en la frontera con Estados Unidos desde el 2011, parte
de los cuales han sido asegurados por miembros de
diferentes agencias tras ser derribados al cruzar hacia
territorio estadounidense…

Who: Mexican cartels; specific cartels not stated. Key hiring cities
for domestic drone production personnel—Mexico City, Querétaro,
Guadalajara, and Nuevo León—suggest Sinaloan and Los Zetas cartel
involvement at a minimum.

What: Unmanned aerial vehicles (UAVs) utilized to transport illicit
narcotics across the border from Mexico to the United States. Initially
foreign produced, particularly Israeli, and then domestic drones since
2014. Since 2012, the Drug Enforcement Administration (DEA) has
documented about 150 of these confirmed UAV intrusions.

When: UAV use since at least 2010 to the present.

Where: Along the Mexican and U.S. border.

Why: For transport of illicit narcotics; profit motive, overcome U.S.
border defenses.

Analysis: The Mexican cartels have engaged in a three phase
evolutionary process of aerial narcotics trafficking from conventional
aircraft (both converted airliners and light craft) to ultralight aircraft
to unmanned aerial vehicles (UAVs). Along the U.S. Southern border,
this process has been prompted by increased U.S. homeland security
activities. UAV use is not unique to the Mexican cartels with such
criminal activity noted in January 2009, with drones going to Elmley
Prison in Sheerness, Kent, United Kingdom (suspected narcotics),[1]

in February 2011 going in to a prison in the Tula region south of Moscow, Russia (heroin),[2] in November 2011 coming across the Straight of Gibraltar into Spain (cannabis resin),[3] throughout 2013 in the province of Quebec, Canada going to various prisons (illicit narcotics),[4] in November 2013 going to the US prison in Calhoun, Georgia (cigarette smuggling),[5] in March 2014 going to a Melbourne, Australia prison (illicit narcotics),[6] and in May 2014 entering the Kaliningrad region, Russia (cigarette smuggling).[7] Of concern are future Mexican cartel UAV evolutionary potentials related to a) sensor payload use for reconnaissance and surveillance functions and b) weapons payload use for small arms and IED attack capabilities. While there are no reports that either of these two evolutionary potentials have taken place, they would provide the Mexican cartels with additional tactical and operational level capabilities.

Significance: Armed Drone Potentials, Cartel Technology, Narcotics Trafficking, Sensor Drone Potentials

Endnotes

[1] Daily Mail Reporter, "Remote control toy helicopter 'used to fly drugs into prison'." *Daily Mail.* 12 January 2009, http://www.dailymail.co.uk/news/article-1112673/Remote-control-toy-helicopter-used-fly-drugs-prison.html.

[2] "Heroin by helicopter." *Voice of Russia.* 1 February 2011, http://voiceofrussia.com/2011/02/01/42369291/.

[3] Tom Worden, "Plane smart! Smugglers use remote-controlled aircraft to fly in cannabis." *Daily Mail.* 21 November 2011, http://www.dailymail.co.uk/news/article-2064276/Plane-smart-Smugglers-use-remote-controlled-aircraft-fly-cannabis.html.

[4] See Michael Aubry, "Drone sets off security alert at Hull jail." *Ottawa Sun.* 26 November 2013, http://www.ottawasun.com/2013/11/26/drone-sets-off-security-alert-at-hull-jail and Brian Anderson, "How Drones Help Smuggle Drugs Into Prison." *Motherboard.* 10 March 2014, http://motherboard.vice.com/read/howdrones-help-smuggle-drugs-into-prison.

[5] "Crooks get creative to smuggle contraband." WALB News 10. 22 & 27 November 2013, http://www.walb.com/story/24047984/crooks-get-creative-to-smuggle-contraband.

[6] Brian Anderson, "How Drones Help Smuggle Drugs Into Prison." *Motherboard.* 10 March 2014, http://motherboard.vice.com/read/how-drones-help-smuggle-drugs-into-prison and Nick Evershed, "Drone used in attempt to smuggle drugs into Melbourne prison, say police." *The Guardian.* 10 March 2014, http://www.theguardian.com/world/2014/mar/10/drone-used-in-attempt-to-smuggle-drugs-into-melbourne-prison-say-police.

[7] "Lithuanian drone detained in Russia last week was used for cigarette smuggling." ITAR-TASS News Agency. 14 May 2014, http://en.itar-tass.com/world/731634.

Further Reading

Marguerite Cawley, "Drone Use in Latin America: Dangers and Opportunities." *Insight Crime.* 18 April 2014, http://www.insightcrime.org/news-analysis/drone-use-in-latin-america-dangers-and-opportunities.

Chapter 3

Mexican Drug Traffickers Using Drones to Bring Drugs into the United States

Brenda Fiegel

Initially Published in OE Watch October 2014
Issue on 17 August 2014, p. 22

Source: "Narco envía droga a EU... en drones." Ejecentral. Accessed on August 17, 2014 from http://www.eluniversal.com.mx/nacion-mexico/2014/carteles- fabrican-narcodrones-trasiego-eu-1022274.html:

Drones Used as Drug Mules to Smuggle Drugs into the United States

Mexican drug cartels are using drones to transport drugs across the US/Mexico border. Authorities first became aware of this phenomenon in 2011 after multiple drones detected by radar systems were shot down upon their entrance into the United States. Since their first detection, two significant changes associated with drone use by cartels have been reported. First, Mexican drug cartels are now placing orders to drone producers in Mexican-based cities including the Federal District, Guadalajara, Monterrey, Querétaro and Tijuana. This is a change because at the start of 2011, virtually all drones owned and operated by cartels were produced abroad; primarily in Israel. Because of the change, Mexican based drone producers are also benefiting as they are paid 2-3 times the asking price for the final product. Second, drug cartels have realized that using drones to transport drugs does not carry as much risk or cost as much as traditional means of transport (human traffickers, narco tunnels, semi-submersibles). Regarding risk, it is difficult to track a drone back to a particular group.

OE Watch Commentary: According to this source article, Mexican cartels have been using drones to transport drugs across the US-Mexican border since around 2011. This same source further indicated that some of the first drones detected by radar systems were shot down in central Arizona. By 2012 drone use along the border had rapidly increased, as evidenced by the US interception of 150 drones carrying an estimated two metric tons of drugs, primarily marijuana. Today, Mexican sources describe drones as being the "perfect drug mule," given that they involve less risk to drug-trafficking organizations and their personnel. Additionally, they cost significantly less than drug tunnels and semi- submersibles, and are even able to transport cash shipments from the US into Mexico without being detected. As the article points out, with the help of Mexican-based drone producers, original models have been modified and are now able to transport anywhere from 60 to 100 kg of drugs. This home- based manufacturing system also benefits Mexican drone producers, as they are paid 2-3 times the asking price for the final product. **End OE Watch Commentary (Fiegel)**

Chapter 4

Border Patrol foils drone drug incursion into the U.S.

U.S. Customs and Border Protection

Initially Published in CBP Newsroom on 12 January 2016 (Release Date) [1]

YUMA, Arizona – The U.S. Customs and Border Protection recovered approximately 30.8 pounds of marijuana dropped from a drone near San Luis, Arizona, Nov. 16, 2015.

This is the first drone drug incursion detected by CBP.

"The highly effective enforcement techniques employed by Yuma Sector and throughout the Border Patrol often forces smuggling organizations to redirect their efforts," says Chief Patrol Agent Anthony J. Porvaznik. "As a result, they experiment with different techniques in an attempt to bring narcotics and other harmful contraband into the United States. This means we must adapt and meet these new challenges. Our agents' vigilance was responsible for detecting this particular drone event, but we always encourage members of the public to assist our efforts by contacting the Border Patrol upon seeing suspicious activity."

Yuma Station Border Patrol Agents observed an OctoCopter style drone illegally enter the San Luis airspace from San Luis, Rio Colorado, Mexico, and jettison a bundle. Aided by night vision goggles, agents were able to follow the drone to its drop point, where three bundles of marijuana - weighing approximately 10 pounds each - were discovered along the bank of a canal. The drugs had an estimated value of $15,430.

The drugs were processed per Yuma Sector guidelines.

Yuma Sector Border Patrol agents effectively combat smuggling organizations attempting to illegally transport people and contraband through southwestern Arizona and California. Citizens can help the Border Patrol and U.S. Customs and Border Protection by calling 1-866-999-8727 toll-free to report suspicious activity. Callers can remain anonymous.

Endnotes

[1] Release Date: 12 January 2016: https://www.cbp.gov/newsroom/local-media-release/border-patrol-foils-drone-drug-incursion-us.

Addendum (Anthology)

Thermal image:

Thermal Image of Cartel Drone (Marijuana Smuggling)
Taken at 10:19:40 PM on 16 November 2015
Source: U.S. Customs and Border Patrol (CBP)
Posted at: http://www.tucsonsentinel.com/local/report/011316
yuma_drone/bp-agents-intercept-drone-dropped-drugs-first-time/.

Chapter 5

Mexican Cartel Strategic Note No. 18: Narcodrones on the Border and Beyond

John P. Sullivan and Robert J. Bunker

Initially Published in Small Wars Journal on 28 March 2016

Drones (also called unmanned aerial vehicles or systems—UAVs or UAS) are becoming more common along the US-Mexico border as Mexican cartel assets. While typically considered a tool for smuggling[1], they are increasingly seen as having application for other purposes including espionage, surveillance, and as weapons.

A recent opinion piece at *Fox News Latino* in February 2016 underscores the emerging potentials of aerial drones on our Southern frontier:

> There are legions of unmanned aerial vehicles (UAVs) patrolling the U.S.-Mexico border, and they do not belong to the United States. They are owned and operated by Mexican drug cartels, and there's nothing we can currently do to stop them.
>
> The proliferation of small, cheap UAVs (aka drones) has raised a litany of security concerns, from interference

with commercial aviation to possible delivery systems for weapons. Along the U.S. southern border, the Mexican cartels are operating drones as intelligence gathering tools.[2]

The author, Nelson Balido a border security analyst, suggests that drones are replacing *halcones* (lookouts and spies) as intelligence gathers along the border. Balido observed that cartels are also using drones for smuggling and intelligence gathering. He notes that the cartels use GPS (Global Positioning System) countermeasures against US aerial drones. He further suggests that the "fleet of drones" weakens US border security efforts and forging an immediate response is essential to countering the proliferation of this threat.[3] In this strategic note, we assess the current threats and emerging potentials posed by drug cartels use of drones along the US-Mexico border and other frontiers. We also consider recent incidents and analyses to consider potentials involving use for criminal enterprise and as weapons systems for violent non-state actors (VNSAs) such as gangs, cartels, terrorists, and private armies (criminal soldiers).

Recent Incidents (Tactical Indicators)

In April 2015 border authorities thwarted an attempt to smuggle 30 pounds of heroin from Baja to Calexico.[4] In a second incident, earlier in January of that same year, a drone carrying about 6 pounds of methamphetamine (*cristal*) crashed in a parking lot near the San Ysidro border crossing near San Diego.[5] The narcodrone found by the Tijuana Municipal Police near the San Ysidro crossing was a commercially available Chinese model known as "Spreading Wings 900." The model is typically used for photography.[6]

Photo of "Spreading Wings 900" Drone (Dron) recovered by Tijuana Municipal Police on 20 January 2015 near San Ysidro border crossing in the Zona del Río, Source: Secretaria de Seguridad Pública Municipal Tijuana, Mexico [For Public Distribution]

Other criminal use of drones has included smuggling contraband into prisons[7] and locating and stealing illicit cannabis grows.[8] As Robert J. Bunker noted in Mexican Cartel Tactical Note # 21 (2014), Mexican drug cartels have been using narcodrones since at least 2010.[9] Bunker assessed the situation at that time as follows:

> The Mexican cartels have engaged in a three phase evolutionary process of aerial narcotics trafficking from conventional aircraft (both converted airliners and light craft) to ultralight aircraft to unmanned aerial vehicles (UAVs). Along the U.S. Southern border, this process has been prompted by increased U.S. homeland security activities. UAV use is not unique to the Mexican cartels with such criminal activity noted in January 2009, with drones going to Elmley Prison in Sheerness, Kent,

United Kingdom (suspected narcotics), in February 2011 going in to a prison in the Tula region south of Moscow, Russia (heroin), in November 2011 coming across the Straight of Gibraltar into Spain (cannabis resin), throughout 2013 in the province of Quebec, Canada going to various prisons (illicit narcotics), in November 2013 going to the US prison in Calhoun, Georgia (cigarette smuggling), in March 2014 going to a Melbourne, Australia prison (illicit narcotics), and in May 2014 entering the Kaliningrad region, Russia (cigarette smuggling). Of concern are future Mexican cartel UAV evolutionary potentials related to a) sensor payload use for reconnaissance and surveillance functions and b) weapons payload use for small arms and IED attack capabilities. While there are no reports that either of these two evolutionary potentials have taken place, they would provide the Mexican cartels with additional tactical and operational level capabilities.[10]

While current statistics on cartel border drone use are unavailable, there were an estimated 150 UAV intrusions across the border from 2012 through 2014.[11]

Bunker's concerns that aerial drones (UAVs) would be used by narco-cartels for intelligence, surveillance, and reconnaissance (ISR) (more recently shared by Badillo) are being realized. Recent reports reinforce perceptions that narcodrones are being used for smuggling and ISR along the US-Mexico frontier. Drug enforcement and border security officials aren't concerned that drone use will become a major smuggling method since their payload is limited. The real concern is the use of aerial drones (UAVs) as observation platforms for reconnaissance and surveillance. The tactical intelligence gathered can be used to select routes for high value, high volume shipments, and for targeting border patrol agents and law enforcement officials.[12]

Counter-Drone Aspects of Mexico's Drone Wars

In addition to ISR, the narcos are employing counter-drone electronic countermeasures to thwart US and Mexican security forces who use drones for interdiction. US drones (including Predators) routinely monitor the border and littoral spaces looking for drug shipments, illegal migration, human traffickers, and other illicit activity. US drones are joined by UAVs operated by Mexico's security services (Federal Police, SEDENA [army], SEMAR [navy], and CISEN [intelligence])[13], local police, and the cartels, making the border's airspace complex and contested.

Drug cartels have been spoofing (and hacking and jamming) US border security drones to divert the security surveillance, allowing uninterrupted cross-border incursions by the narcos.[14] Such EW [Electronic Warfare] activity focusing on the disruption of law enforcement C² (command and control) and communications activities is a national evolution of earlier cartel operations going back to at least as early as 2008.[15]

It is not inconceivable that narcos could up-gun aerial drones (as well as land mobile unmanned vehicles and vessels) to include ballistic, laser, or explosive (IED) capabilities.[16] Recognizing the difficulty of countering small narcodrones with conventional means, some police agencies are training eagles and falcons to interdict small hostile drones.[17] Additionally, other agencies are looking into using their own drones with trailing nets, kinetic drone-on-drone kill options, and even officers with skeet shooting skills for special anti-drone assignment to address venue and special event protection needs.

Assessing the Narcodrone Threat

Drones have been used by narcos and gangs for smuggling and for intelligence (ISR). In addition, non-state actors (insurgent, criminal, corporate, and activists) have used drones to stage IED-type attacks and gather intelligence. These actors have included individuals (lone wolfs), terrorists and insurgents (including Hezbollah and Islamic State (IS), and organized crime groups (cartels) in Mexico and Colombia. These unmanned drones have included UAVs—ranging from scale model jet aircraft through smaller hobbyist and gray area systems into specialized micro-infiltrator airframes—as well as unmanned semi-submersibles, and remotely-piloted ground vehicles. Corporations and activists are likely to use drones for security, publicity, and intelligence gathering purposes.[18]

Security services are concerned about drones because they can be used for a range of purposes from smuggling high value commodities like heroin, Ecstasy, and methamphetamines to their ISR uses, to being used as weapons platforms for explosive, ballistic, laser, and biological weapons.[19]

The Remote Control Project at the Oxford Research Group conducted an assessment of commercially-available unmanned vehicles (drones) including unmanned aerial vehicles (UAVs), unmanned ground vehicles (UGVs) and unmanned marine vehicles (UMVs) (vessels including surface and submersible variants). Their analysis also summarized the range of potential non-state actors using drones: lone wolves, terrorist organizations, insurgent groups, organized crime groups, corporations, and activist groups. They viewed the target sets as involving both domestic and international theatres and involving long-term static targets such as embassies and critical facilities (such as nuclear power stations), temporary static targets such as G7 summits or political events, and mobile targets such as convoys or vehicles. They also emphasized the need to develop countermeasures for both attack and ISR operations. Drone countermeasures (or C-UAS [Counter Unmanned Aerial Systems] in military parlance) range from regulatory measures restricting access and use through passive countermeasures (early warning, signal jamming and takeover) to active kinetic and laser defense systems.[20]

The Open Briefing group also raised the possibility of Islamic State (IS) drone threats since IS has used aerial drones for battlefield intelligence in the Iraqi/Syrian theatre and has attempted to use both aerial and ground drones with explosive payloads (IEDs) to attack Kurdish forces.[21]

Pirates should also be added to the list of actors suggested above. A recent Canadian assessment points out that criminals at sea (pirates or maritime non-state actors-MNSAs) are using GPS and bootleg submarines to operate despite the presence of naval forces. These pirates, smugglers, or traffickers could use aerial drones to gain air capacity for surveillance and weapons delivery.[22]

Esther Kersley, in her essay "Drones, drugs and death," observes that 'remote control' warfare, involving mass surveillance, drones, military special forces, and private military and security companies (PMSCs) is on the rise.[23] Kersey notes that both armed and reconnaissance drones are employed by states (notably the US) and that drones are increasingly being found in the hands of transnational criminal organizations (i.e. drug cartels). Thus, remote warfare technology is being used by cartels to smuggle contraband and "to fight each other, dominate criminal markets, control local populations and deliver lethal action against their enemies."[24]

Vanda Felbab-Brown in February 2016 has now contributed her analysis. She posits that criminal groups (gangs and cartels) are embracing remote lethal action to counter state law enforcement and security measures. The remote fighting capabilities suggested by Felbab-Brown include surveillance cameras and drones. Drones are beginning to be used not only for transporting illicit goods but also, more importantly, as weapons:

> The new radical remote-warfare development on the horizon is for criminal groups to start using drones and other remote platforms not merely to smuggle and distribute contraband, as they are starting to do already, but to deliver lethal action against their enemies— whether government officials, law enforcement forces, or rival crime groups.[25]

Of course, countermeasures and drone vs. drone combat are a likely consequence of gangsters and mafias embracing remote warfare tactics.

Drones (now aerial platforms and soon land and water-borne variants) are changing the shape of non-state conflict. Drug, crime, and gang wars are likely to include the use of drones as both contraband transport and ISR platforms in the near-term. As the technology becomes more mature and accessible, logistical models and lethal variants (kinetic, explosive, and laser) are likely to be exploited as effective force sustainment and weapons systems in both urban and rural settings.

Conclusion: Criminal Drones

Drones were initially considered a high technology, high expense capability available to national forces. They were also viewed as a component of the more recognized cruise missile threat. As technology matured, commercial-off-the-shelf (COTS), and do-it-yourself (DIY) capabilities have emerged which have created a stand-alone threat category. These systems are readily available to narcos, terrorists, and insurgents. Individual drones, and swarms of drone packs are probable features of narcotrafficking and crime wars. These robotic actors are likely to involve airborne, land, and waterborne (riverine, littoral, and maritime) variants on all sides of the confrontation (military, police and security services, the media, and criminal, terrorist, and insurgent actors).

The *plazas, colonias, favelas,* and megacities of the drug war zone will become the operational space of future state vs. non-state conflict (embracing local high intensity crime through criminal insurgency). The use of drones is no longer merely a tactical issue; it has strategic and operational potentials for states and their competitors. This threat will become especially acute when larger packs and swarms of semi-autonomous and autonomous armed drones—eventually custom printed to maximize specific mission requirements—are employed utilizing network C^2 architectures that allow for collective decision making protocols and strategies to be implemented.

Endnotes

[1] "Los narcodrones de la frontera," *Telemundo Local*, 3 February 2015 at http://www.telemundo51.com/noticias/Los-narcodrones-de-la-frontera-narcotrafico-nogales-arizona-mexico-eeuu-290686221.html.

[2] Nelson Balido, "Nelson Balido: Mexican cartels patrol border with drones – and U.S. has no response," *Fox News Latino*, 19 February 2016 at http://latino.foxnews.com/latino/opinion/2016/02/19/nelson-balido-mexican-cartels-patrol-border-with-drones-and-us-has-no-response/.

[3] Ibid.

[4] Associated Press, "Heroin smugglers turn to drones," *The Telegraph*, 13 August 2015 at http://www.telegraph.co.uk/news/worldnews/northamerica/usa/11800080/Heroin-smugglers-turn-to-drones.html; and Sarah Berger, "Mexico Drug Trafficking: Drone Carries 28 Pounds of Heroin Across Border To US," *International Business Times*, 13 August 2015 at http://www.ibtimes.com/mexico-drug-trafficking-drone-carries-28-pounds-heroin-across-border-us-2051941.

[5] Arturo Salinas y Manuel Ocaño, "FOTOGALERÍA: Cae dron que transportaba droga en Tijuana,' *Excelsior*, 22 January 2015 at http://www.excelsior.com.mx/nacional/2015/01/22/1003922; Rob Crilly, "Meth-laden drone crashes near US-Mexico border," *The Telegraph*, 22 January 2015 at http://www.telegraph.co.uk/news/worldnews/centralamericaandthecaribbean/mexico/11361825/Meth-a-drone-crashes-near-Mexican-border.html; Rafa Fernandez De Castro, "Meth from Heaven? Narco drone falls out of Tijuana sky,' *Fusion*, 22 January 2015 at http://fusion.net/story/39975/meth-from-heaven-narco-drone-falls-out-of-tijuana-sky/.

[6] Said Betanzos, "Cayó 'narcodrón' cerca de garita," *El Mexicano*, 22 January 2015 at http://www.el-mexicano.com.mx/informacion/noticias/1/3/estatal/2015/01/22/819263/cayo-narcodron-cerca-de-garita; and "SSPM REPORTA FORMAS DE TRASIEGO," Secretaría de Seguridad Pública Tijuana" at https://www.facebook.com/policiatijuana/posts/829170020509500.

[7] David Francis, "Want to Smuggle Drugs into Prison? Buy a Drone" *Foreign Policy*, 4 August 2015 at http://foreignpolicy.com/2015/08/04/want-to-smuggle-drugs-into-prison-buy-a-drone/; and Justin Ling, "Someone Used a Drone to Deliver a Handgun Into a Notorious Canadian Prison," *Vice*, 14 December 2015 at https://news.vice.com/article/someone-used-a-drone-to-deliver-a-handgun-into-a-notorious-canadian-prison.

[8] Drones equipped with cameras, heat sensors, and Wi-Fi have been employed by gangsters in the UK to case and then raid marijuana farms. Matt Whitlock, 'English Gangs are now using Drones to Locate and Steal From, Weed Farms," *The Source*, 24 April 2014 at http://thesource.com/2014/04/24/english-gangs-are-now-using-drones-to-locate-and-steal-from-weed-farms/.

[9]   Robert J. Bunker, "Mexican Cartel Tactical Note #21: Cartel Unmanned Aerial Vehicles (UAVs)," *Small Wars Journal*, 1 August 2014 at http://smallwarsjournal. com/blog/mexican-cartel-tactical-note-21.

[10] Ibid. Citations are available at the original article.

[11] Ibid. Also see "'Narcodrones', la nueva técnica de los cárteles mexicanos," *El Comercio*, 16 July 2014 at http://elcomercio.pe/mundo/latinoamerica/ narcodrones-nueva-tecnica-carteles-mexicanos-noticia-1743520; and Doris Gómora, "Fabrican narcos sus propios drones, alerta la DEA," *El Universal*, 9 July 2014 at http://archivo.eluniversal.com.mx/nacion-mexico/2014/carteles-fabrican-narcodrones-trasiego-eu-1022274.html.

[12] See Camilo Mejia Giraldo, "Mexico's Cartels Building Custom-Made Narco Drones: DEA," *InSight Crime*, 11 July 2014 at http://www.insightcrime.org/ news-briefs/mexico-s-cartels-building-custom-made-narco-drones-dea; and Andrew O'Reilly, "DEA: Narco-drones not major smuggling concern, but could help set up attacks on agents," *Fox News Latino*, 22 January 2015 at http://latino. foxnews.com/latino/news/2015/01/22/dea-narco-drones-not-major-smuggling-concern-but-could-help-set-up-attacks-on/.

[13] Jan-Albert Hootsen, "Inside Mexico's Drone Wars," *Voacative*, 6 January 2014 at http://www.vocativ.com/world/mexico-world/inside-mexicos-drone-wars/.

[14] See Waqas, "US Border Patrol Drones Hacked by Drug Cartels," *Hackread*, 3 January 2016 at https://www.hackread.com/us-border-patrol-drones-hacked-by-drug-cartels/; Matthew H. Fleming, et al, "Unmanned Systems in Homeland Security," Homeland Security Studies and Analysis Institute and Center for Strategic and International Studies, January 2015 at http:// csis.org/files/attachments/151216 Unmanned Systems.pdf; and Patrick Tucker, "DHS: Drug Traffickers Are Spoofing Border Drones," *Defense One*, 17 December 2015 at http://www.defenseone.com/technology/2015/12/ DHS-Drug-Traffickers-Spoofing-Border-Drones/124613/.

[15] While not all the LE communications disruption taking place is intentional, a specific component of it is. See Diana Washington Valdez, "Cartels, Mexican army blamed for interference." *El Paso Times*. 18 August 2008. For a mirror of this article see http://news.wexico.com/technology/00aug2008/mexinterference. htm. Also see "TSG IntelBrief: Mexican Drug Cartels in the Cyber Age." *The Soufan Group*. 13 November 2012 at http://soufangroup.com/tsg-intelbrief-mexican-drug-cartels-in-the-cyber-age/ and Patrick Tucker, "DHS: Drug Traffickers Are Spoofing Border Drones." *Defense One*. 17 December 2015 at http://www.govexec.com/defense/2015/12/homeland-security-drug-traffickers-are-spoofing-border-drones/124617/.

[16] For a discussion of IED drones see Robert J. Bunker, "Virtual Martyrs: Jihadists, Oculus Rift, and IED Drones," *TRENDS Research & Advisory*, 14 December 2014 at http://trendsinstitution.org/?p=762.

[17] Anthony Cuthbertson, "Police Train Eagles to Take Out Rogue Drones," *Newsweek*, 2 February 2016 at http://www.newsweek.com/police-train-eagles-take-out-rogue-drones-422030drones; and Patrick Barkham, "Drone-fighting eagles – a reminder of nature's superpowers," *The Guardian*, 8 February 2016 at http://www.theguardian.com/commentisfree/2016/feb/08/drone-fighting-eagles.

[18] Chris Abbott, Matthew Clarke, Steve Hathorn, and Scott Hickie, "An Assessment of Known Drone Use by Non-State Actors," *ISN, ETH Zurich*, 26 January 2016 at http://www.isn.ethz.ch/Digital-Library/Articles/Detail/?id=195707.

[19] Dane Schiller, "Be very afraid of drones, warns expert," *Houston Chronicle (Chron)*, 26 January 2016 at http://blog.chron.com/narcoconfidential/2015/01/be-very-afraid-of-drones-warns-expert/; and Dean Klovens, "UAVs: The future of terrorist weaponry?" *Global Risks Insight*, 1 February 2016 at http://globalriskinsights.com/2016/02/uavs-the-future-of-terrorist-weaponry/?utm_content=buffer72736&utm_medium=social&utm_source=twitter.com&utm_campaign=buffer.

[20] Chris Abbott, Matthew Clarke, Steve Hathorn, and Scott Hickie, "Hostile Drones: The Hostile Use of Drones by Non-State Actors Against British Targets." *Remote Control Project, Oxford Research Group, and Open Briefing*, January 2016 at http://remotecontrolproject.org/wp-content/uploads/2016/01/Hostile-use-of-drones-report_open-briefing_16.pdf.

[21] Chris Abbott and Matthew Clarke, "How to respond to the threat from hostile drones in the UK," *Open Briefing*, 14 March 2016 at http://www.openbriefing.org/thinktank/publications/how-to-respond-to-the-threat-from-hostile-drones-in-the-uk/.

[22] Jordan Pearson, "Drones Will Forever Change How Pirates Operate on the High Seas," *Motherboard*, 11 February 2016 at http://motherboard.vice.com/read/drones-will-forever-change-how-pirates-operate-on-the-high-seas-canada.

[23] Esther Kersley, "Drones, drugs and death," *openDemocracy,* 17 March 2016 at https://www.opendemocracy.net/drugpolicy/esther-kersley/drones-drugs-and-death. The interplay between private military actors and criminals discussed in Kersey has also been addressed by the current authors, see especially John P. Sullivan, "Terrorism, Crime and Private Armies," *Low Intensity Conflict & Law Enforcement*, Volume 11, Issue 2-3, 2002, pp. 239-253; and John P. Sullivan and Robert J. Bunker, "Drug Cartels, Street Gangs, and Warlords," *Small Wars & Insurgencies*, Volume 13, Issue 2, 2002, pp. 40-53.

[24] Ibid (Kersley). Kersley's essay is also republished as Esther Kersley, "New Tactics, Old Strategy: Remote Warfare and the War on Drugs," *Oxford Research Group*, 18 March 2016 at http://www.oxfordresearchgroup.org.uk/publications/briefing_papers_and_reports/new_tactics_old_strategy_remote_warfare_and_war_drugs.

[25] Vanda Felbab-Brown, "Drugs and Drones: The Criminal Empire Strikes Back," *Remote Control Project Blog*, 24 February 2016 at http://remotecontrolprojectblog.org/2016/02/24/drugs-and-drones-the-crime-empire-strikes-back/.

Additional Reading

Chris Abbott, Matthew Clarke, Steve Hathorn, and Scott Hickie, "Hostile Drones: The Hostile Use of Drones by Non-State Actors Against British Targets." *Remote Control Project, Oxford Research Group*, and *Open Briefing*, January 2016 at http://remotecontrolproject.org/wp-content/uploads/2016/01/Hostile-use-of-drones-report_open-briefing_16.pdf.

Robert J. Bunker, *Terrorist and Insurgent Unmanned Aerial Vehicles: Use, Potentials, and Military Implications*. Carlisle Barracks: US Army War College, Strategic Studies Institute, August 2015 at http://www.strategicstudiesinstitute.army.mil/pubs/display.cfm?pubID=1287.

Marguerite Cawley, "Drone Use in Latin America: Dangers and Opportunities," *InSight Crime*, 18 April 2014 at http://www.insightcrime.org/news-analysis/drone-use-in-latin-america-dangers-and-opportunities.

Vanda Felbab-Brown, "Drugs and Drones: The Criminal Empire Strikes Back," *Remote Control Project Blog*, 24 February 2016 at http://remotecontrolprojectblog.org/2016/02/24/drugs-and-drones-the-crime-empire-strikes-back/.

Chapter 6

Yuma Border Patrol Experiencing Drone Activity

U.S. Customs and Border Protection

Initially Published in CBP Newsroom on 6 April 2016 (Release Date) [1]

YUMA, Ariz. – The Yuma Sector Border Patrol has recently encountered small remote controlled aircraft, commonly referred to as drones, being used to smuggle drugs into the United States. The drones vary in size, but are commonly between 2 to 4 feet wide.

Drones have been observed primarily in the San Luis area. They are known to carry illegal contraband into the U.S. where it is dropped and picked up by smugglers north of the border.

Yuma Border Patrol agents are experiencing recent activity involving remote controlled aircraft, or drones

Endnotes

[1]   Release Date: 6 April 2016 https://www.cbp.gov/newsroom/local-media-release/ yuma-border-patrol-experiencing-drone-activity?_ga=2.155386559.58479 5156.1631411989-349536333.1631411989.

Chapter 7

Narco-Drones: A New Way to Transport Drugs

Brenda Fiegel

Initially Published in Small War Journal on 5 July 2017

"A narcotrafficking technique first used in Mexico now expands to other countries in Central and South America."

In mid-November, Colombian Police seized 130 kilograms of cocaine and a drone used by narcotraffickers in the Bahía Solano sector of Chocó, allegedly used to send cocaine shipments to Panama. This information is of interest as it is the first instance in which drones have been identified as a viable trafficking method in the country, according to Colombian news source *La Prensa*.[1]

The drones being utilized can transport up to 10 kilograms (22 pounds) of cocaine and travel up to 100 kilometers (62 miles) in a single trip. Authorities also indicated that this method was likely developed by the Clan del Golfo (formerly known as Clan Úsuga) which is the largest criminal gang in Colombia dedicated to drug trafficking.[2]

The use of this tactic by the Clan del Golfo is in no way surprising as they are known for their diverse trafficking methods, which range from using illegal migrants to carry small shipments through jungle regions into Panama, to submarines that can travel all the way to the United States.[3]

However, it is significant as it demonstrates the capabilities of cartels in adjusting their trafficking methods to changing operational conditions meaning that no matter what obstacles are placed in front of them, drug trafficking organizations will overcome virtually any barrier if humanly possible to move their product and generate profits. This means that authorities must be vigilant in anticipating trafficking changes because as soon as one method is blocked or temporarily unusable, another will quickly be found to replace it.

Mexican Tactics

While news regarding drug cartels using drones in Colombia is a new phenomenon, this tactic has been used by Mexican cartels since around 2010, and it is likely that Mexican success motivated the Clan del Golfo to test the same method in their territory. In fact, by 2012, drone use along the border was highly prevalent as evidenced by the United States' interception of 150 drones carrying an estimated two metric tons of drugs; primarily marijuana, cocaine, and heroin.[4]

In fact, Mexican cartels have become so vested in drone use that they are now using Mexican-based companies to produce them in cities including the Federal District, Guadalajara, Monterrey, Querétaro and Tijuana. This is interesting because prior to 2011, virtually all drones owned and operated by cartels were produced abroad; primarily in Israel and China.

The new Mexican-made drones are very different from the ones used for personal use as they can supposedly transport anywhere from 60-100 kilograms (132-220 lbs.) of drugs in a single trip, and it is likely that engineers will continue working to make trafficking drones more efficient in terms of weight they can carry, distance they can travel, and methods to deter their detection.[5]

In terms of current use, drones used to transport drugs usually operate during the night, and never even land on U.S. soil. They simply drop the shipment and return to Mexico.

The Perfect Drug Mule

It can be surmised that drones are gaining popularity in both Mexico and Colombia as they represent a nearly perfect drug mule in the sense that they involve less risk to drug trafficking organizations and their employees who represent possible risks to any cartel if the arrested individual provides authorities regarding TTPs, routes, shipment time tables, etc. Additionally, drones, in comparison to their human counterparts cost significantly less as a drug mule can earn as much as $10,000 for successful delivery of a single shipment. However, when compared to other non-human forms of transport, the cost of drones is simply insignificant when comparing them to the cost of building drug tunnels (Mexican tactic), semi-submersibles (universal DTO tactic), and submarines (Colombian tactic).

Drones only apparent flaws are that they are not capable of traveling long distances or carrying large-scale shipments at this point. Regardless, it is likely that drone use by drug cartels in Mexico and Colombia will increase; especially if producers work on developing more agile models that can carry added weight and fly longer distances at lower altitudes.

Other aspects that make drones appealing to cartels is that their use is multi-faceted and does not necessarily pertain to just drug trafficking activities. For example, drones can conduct surveillance operations, perform intelligence gathering, and even move money or valuable messages short distances without being detected. In fact, drone use for counter-surveillance purposes was reported in June 2017 in Australia where an unidentified DTO utilized drones to surveil police activity in an attempt to protect a $30,000,000 cocaine shipment that originated in Panama and was likely of Colombian origin. This is not to say that the Colombians were responsible for informing their contacts in Australia to use drones for counter-surveillance, but the possibility certainly exists based on what is known about drone use by South American cartels.[6]

The Future of Drones in Mexico and South America

When comparing borders, Mexico has a definite advantage in terms of drone use strictly for drug trafficking purposes based on the following three reasons. First, the Southwest border is vast and highly difficult to monitor. Second, in many cases drones are moving from one populated area, directly into another, which helps camouflage their activity. Third, drones departing from Mexico to the U.S. are travelling much shorter distances than those departing from Colombia to Panama, meaning there is less risk of detection and malfunction of the equipment.

As for drones in South America, border security issues have always existed between Panama and Colombia because their shared geography is a hot spot for drug, human, and weapons trafficking, as well as money laundering. More specifically, the so-called "Darién Gap" (classified by authorities from both countries as a lawless jungle region along the two borders) serves as an epicenter of illegal activity because it is completely under the control of drug trafficking organizations the Revolutionary Armed Forces of Colombia (FARC).

As neighbors, Panama and Colombia have historically worked together to combat the aforementioned issues, but are now looking to further this cooperation. As part of their commitment to improve border security, Panamanian President Juan Carlos Varela and Colombian President Juan Manuel Santos stated that they plan to install two security bases in La Olla and La Balsa, in addition to merging two other security outposts in Alto Limón and La Unión.

All of these bases are located in the Darién Region, and will be staffed with both Colombian and Panamanian security forces. They are expected to be operational at some point in 2017. As indicated by President Santos, this collaboration is expected to have a positive impact in decreasing all types of illegal activity in the region, and will possibly thwart the new imposition of drone use in the area before it gains the same popularity it has in Mexico.

Endnotes

[1] "Narcotraficantes envian cocaina a Panama con drones: Policia de Colombia (Colombian Drug Traffickers Send Cocaine to Panama with Drones)," *La Prensa*, 17 November 2016. http://www.prensa.com/mundo/Narcotraficantes-enviando-Panama-Policia-Colombia_0_4622537754.html.

[2] "Descubren un 'narcodron' en Colombia; enviaba cocaína a Panamá (Authorities Discover Narco-Drone in Colombia with Cocaine Shipment Headed to Panama)," *Excelsior*, 15 November 2016. http://www.seguridadenamerica.com.mx/noticias/de-consulta/secciones-revist-seguridad-en-america/noticias-de-control-de-acceso/24684-descubren-un-narcodron-en-colombia-enviaba-cocaina-a-panama.

[3] "Policía revela 'el hormigueo', nueva modalidad del Clan Úsuga para sacar droga del país (Police Reveal New Micro-Trafficking Scheme Utilized by Clan Úsuga to Move Drugs from Colombia to Panama)," *Noticias CMI*, 10 October 2016. http://www.cmi.com.co/policia-revela-el-hormigueo-nueva-modalidad-del-clan-Úsuga-para-sacar-droga-del-pais.

[4] "Narco envía droga a EU… en drones (Drug Traffickers Use Drones to Ship Drugs to the United States)," *Ejecentral*, 17 August 2014. http://www.eluniversal.com.mx/nacion-mexico/2014/carteles-fabrican-narcodrones-trasiego-eu-1022274.html.

[5] "Fabrican narcos sus propios drones, alerta la DEA (The DEA Reports that Narcos are Building their Own Drones)," *La Nacion*, 09 June 2014. http://archivo.eluniversal.com.mx/nacion-mexico/2014/carteles-fabrican-narcodrones-trasiego-eu-1022274.html.

[6] "Drug ring 'used drones to counter police' before $30m haul seized in Melbourne: AFP," *ABC.net*, 30 June 2017. http://www.abc.net.au/news/2017-06-30/seven-arrested-in-melbourne-over-international-drug-syndicate/8666176.

Chapter 8

Smuggler Using Drone Busted by Border Patrol

U.S. Customs and Border Protection

Initially Published in CBP Newsroom on 18 August 2017 (Release Date)[1]

SAN DIEGO—A 25-year-old man in possession of several pounds of methamphetamine was arrested by Border Patrol agents after a remote-controlled drone was observed flying over the border last week.

Drone suspected of hauling 12 packages of meth.

On August 8, at about 11:25 p.m., a Border Patrol agent observed a drone fly over the border fence at an area approximately two miles west of the San Ysidro Port of Entry. The agent notified other agents in the area to be on the lookout for the drone.

An agent on an all-terrain vehicle spotted a male suspect at about 11:40 p.m. near the border at Servano Avenue and Valentino Street. The agent approached the man and discovered that he was carrying a large open bag that had a multiple plastic-wrapped packages containing methamphetamine.

After the agent arrested the man, a search of the immediate area was conducted, leading to the discovery of a drone[2] that was concealed under a bush. The drone was approximately two feet in height.

"Due to the agents' heightened vigilance, this drone smuggling scheme was stopped before these dangerous narcotics could enter our communities," said Acting Assistant Chief Patrol Agent Boone Smith.

The methamphetamine had a total weight of 13.44 pounds and an estimated street value of $46,000.

To prevent the illicit smuggling of humans, drugs, and other contraband, the U.S. Border Patrol maintains a high level of vigilance on corridors of egress away from our Nation's borders. To report suspicious activity to the U.S. Border Patrol, contact San Diego Sector[3] at (619) 498-9900.

Endnotes

[1] Release date 18 August 201,: https://www.cbp.gov/newsroom/local-media-release/smuggler-using-drone-busted-border-patrol.

[2] "MATRICE 600 PRO." DJI. https://www.dji.com/matrice600-pro/info.

[3] "San Diego Sector California." CBP Border Stations. https://www.cbp.gov/border-security/along-us-borders/border-patrol-sectors/san-diego-sector-california.

Addendum

See follow-on information concerning the incident: "Drone Drug Smuggler Gets 12-Year Sentence." *CBP Newsroom*. 31 July 2018, https://www.cbp.gov/newsroom/national-media-release/drone-drug-smuggler-gets-12-year-sentence.

Chapter 9

Mexican Cartel Tactical Note #35: Weaponized Drone/UAV/UAS Seized in Valtierrilla, Guanajuato with Remote Detonation IED ('Papa Bomba') Payload

Robert J. Bunker and John P. Sullivan

Initially Published in Small Wars Journal on 23 November 2017

An improvised explosive drone ('*dron bomba*') was interdicted by Mexican Federal Police/Policía Federal (PF) in Guanajunto in Central Mexico at daybreak of Friday, 20 October 2017. Four men were arrested following a 'high-risk' vehicle stop on the Salamanca-Morelia highway. The discovery of an improvised Unmanned Aerial System (UAS)—also known as an Unmanned Aerial Vehicle (UAV)—is the latest example of cartel weapons evolution. The IED is consistent with recent *papas bombas* (potato bombs) employment by the Cártel Jalisco Nueva Generación (CJNG). Guanajunto is currently contested by several cartels including the CJNG, Los Zetas, and the Sinaloa cartel.

Key Information: "'Dron bomba' listo para detonar a distancia." *AM.* 20 October 2017, https://www.am.com.mx/2017/10/20/sucesos/dron-bomba-listo-para-detonar-a-distancia-385808:

> (REDACCIÓN. SALAMANCA, GUANAJUATO.)
> Los cargos que podrían tener los detenidos pueden llegar a ser por terrorismo.
> *See the 0:59 minute video embedded in this article at "Hallaron gran carga explosiva con detonador a distancia" and Interceptan ¡dron bomba!*
> El Procurador de Justicia del Estado, <u>Carlos Zamarripa Aguirre</u>, confirmó que el <u>dron asegurado a cuatro hombres cerca del municipio de Salamanca contenía una gran carga explosiva con un detonador a distancia.</u>
> El hecho, que ha sido único en Guanajuato, ocurrió la madrugada de hoy en la autopista de cuota Salamanca a Morelia donde elementos de la Policía Federal Preventiva detuvieron a cuatro hombres que tripulaban una camioneta y dentro del vehículo los policías federales encontraron el dron y un arma larga AK47.
> El Procurador de Justicia del Estado, dijo que tras ser analizado el dron, se comprobó que el aparato contenía una cantidad importante de explosivo y que está preparado para ser detonado a distancia.
> Los cargos que podrían tener los detenidos pueden llegar a ser por terrorismo.
> Las autoridades ministeriales investigan si los detenidos pertenecen a un grupo criminal dijo el Procurador.

Key Information: "¿Tomar video? Dron con fines terroristas." *AM.* 21 October 2017, https://www.am.com.mx/2017/10/21/local/tomar-video-dron-con-fines-terroristas--386171:

> Entre los usos de un dron, están para entregar una pizza, ser como un taxi, pero también hay usos maléficos.

Drones con explosivos han sido utilizados por el grupo terrorista Estado Islámico (Isis).

En octubre del año pasado en el norte de Irak las fuerzas kurdas que peleaban contra el grupo terrorista lograron derribar un pequeño dron de cuatro rotores.

Inicialmente pensaron que se trataba de uno de los tantos enviados por Isis para monitorear las zonas de guerra y reconocer las áreas.

Sin embargo, cuando comenzaron a revisarlo el artefacto estalló y mató a dos combatientes.

Se informó entonces que al aparato había sido activado mediante una señal de radiofrecuencia, un mecanismo similar al que tenía el dron descubierto ayer, según informó el procurador Carlos Zamarripa.

Fotos de militantes del grupo extremista mostraban a sus combatientes con drones comunes, como los que se pueden comprar en tiendas e Internet, pero con la carga explosiva incorporada.

Ayer en un hecho inédito en Guanajuato, policías federales detuvieron a cuatro hombres que <u>llevaban un droncon un explosivo montado.</u>

"Se trata de un dron... me acaban de corroborar hace un momento que se trata de un artefacto explosivo, con un detonador remoto y una gran carga explosiva", confirmó el procurador del Estado Carlos Zamarripa Aguirre.

El descubrimiento se produjo en la carretera de cuota Salamanca-Morelia, cerca de la comunidad Valtierrilla.

Key Information: "Viajaban en auto robado: les hallan un dron y explosivos." *El Debate.* 21 October 2017, https://www.debate.com.mx/mexico/Viajaban-en-auto-robado-les-hallan-un-dron-y-explosivos-20171021-0007.html:

Cuatro sujetos fueron detenidos en Salamanca, además traían una cuerno de chivo y un detonador para el explosivo; los podrían acusar de 'terrorismo'

**Los sujetos traían un 'cuerno de chivo' y explosivos.
Foto: Policía Federal**

Guanajuato (El Universal).- Elementos de la Policía Federal capturaron en Salamanca a cuatro hombres que tenían un dron con una carga de material explosivo y con detonador remoto, así como un arma larga conocida como "cuerno de chivo", informó el Procurador General de Justicia del Estado, Carlos Zamarripa.

Los sujetos fueron detenidos la madrugada de este viernes a bordo de un camioneta Mazda con reporte de robo, cuando circulaban en la autopista de cuota Salamanca-Morelia.

Foto: Policía Federal

Los agentes federales precisaron que fueron trasladados al Cereso de Irapuato y puestos a disposición del Ministerio Público, bajo la identificación de Christian N., Ángel N., Eduardo N. y Marcos N.

Zamarripa Aguirre informó que la fiscalía estatal realiza las investigaciones en contra de los detenidos por el vehículo reportado como robado, a efecto de determinar la procedencia de la camioneta encontrada.

Explicó que se hará un desglose al Ministerio Público de la Procuraduría General de la República por el dron y el arma con que se les encontró, pues son de uso exclusivo del Ejército y constituyen delitos de competencia federal.

Key Information: "Dron explosivo: Último artefacto del crimen organizado en México." *HispanTV.* 21 October 2017, http://www.hispantv.com/noticias/mexico/357219/incautan-dron-crimen-organizado-violencia:

Autoridades mexicanas incautaron el viernes un dron con explosivos a cuatro delincuentes presuntamente vinculados al crimen organizado en Guanajuato (centro).

Elementos de la Policía Federal mexicana capturaron en Guanajuato a cuatro hombres que tenían un dron con una carga de material explosivo y con detonador remoto, así como un arma larga conocida como "cuerno de chivo", informó el Procurador General de Justicia del Estado de Guanajuato, Carlos Zamarripa.

Según el funcionario, en otras ocasiones ya se han detectado artefactos explosivos de este tipo, pero nunca instalados sobre un dron. Las autoridades no han podido investigar el lugar en el que los criminales pretendían sobrevolar el dron de marca Fly 3DR.

"Tanto el explosivo como el arma son de uso exclusivo del Ejército", por lo que se hará lo necesario para que los acusados sean procesados "en consecuencia", comentó Zamarripa.

Los detenidos están presuntamente vinculados al crimen organizado y se trasladaban a bordo de un vehículo robado muy cerca del municipio de Salamanca, en una carretera que conecta con el vecino estado de Michoacán, informaron los agentes federales desplegados en la zona.

Zamarripa informó que la fiscalía estatal realiza las investigaciones en contra de los detenidos por el vehículo reportado como robado, a efecto de determinar la procedencia de la camioneta encontrada.

Desde principios de año, los episodios de violencia ligados al crimen organizado se han venido multiplicando en Guanajuato, estado que no hace mucho era considerado entre los más apacibles de México.

La semana pasada, varios cuerpos aparecieron desmembrados en el municipio de Celaya (un modus operandi clásico de sicarios de cárteles), mientras que las autoridades arrestaron a numerosos cabecillas de bandas criminales.

Guanajuato es disputado por los poderosos cárteles Jalisco Nueva Generación, Los Zetas y Sinaloa.

Who: Four individuals (Public Ministry IDs: Christian N., Angel N., Eduardo, N., and Marcos, N.) were arrested by Mexican Federal Police/ Policía Federal (PF) in a stolen vehicle in what would be equivalent to a 'felony' or 'high-risk' stop in the United States.

What: In the back of the stolen vehicle inside the hatchback/rear cargo area, a 3DR Solo Quadcopter in an open case with an IED (*'Papa Bomba'*) attached to it with a sling rope and a remote RF detonator was seized. An AK-47 variant assault rifle, 2 magazines, 13 7.62mm bullets, 3 smart phones, 1 texting phone, and 3 black caps (1 with Guanajuato on it) were also recovered.

When: The stolen vehicle was pulled over at dawn by Federal Police on Friday 20 October 2017.

Where: Near the community of Valtierrilla, along the Salamanca-Moreila highway, in the state of Guanajuato which is about 304 km (189 miles) northwest of Mexico City via roadways.

Why: A weaponized drone/unmanned aerial vehicle (UAV)/unmanned aerial system (UAS) with a remotely detonated IED allows for a precision strike to take place against an intended target. This form of IED can be utilized against a point target for assassination purposes or against an area target—such as a grouping of individuals—for anti-personnel purposes. In addition to using such a drone/UAS for assassination or tactical action purposes, it also offers 'narco terrorism' and/or 'narco insurgency' potentials depending on the intent of its use.

Analysis: This significant incident represents the crossing of a Mexican cartel technology and TTP (tactic, technique, and procedure) use firebreak with drone/UAV/UAS 'weaponization' now taking place. It suggests additional cause for heighted concern related to the evolving Mexican cartel security environment. This weaponized drone seizure follows a human shield use incident taking place in Palmarito, Puebla in May 2017 linked to an armed criminal group[1] and a recent national homicide report in September with 2,564 homicides taking place, putting this year on track to be the most deadly ever (21,200 so far) recorded in Mexico during its ongoing criminal insurgency.[2]

The IED drone recovered is indicative of the fusion of two recent technology use trends taking place within cartel groups in Mexico. The first trend is the use of drones/UAS for primarily narcotics smuggling purposes. The cartels have been using drones since 2010, with possibly 150 UAS intrusions into the U.S. between 2012 and 2014 according to an unconfirmed Drug Enforcement Administration (DEA) report. In addition to smuggling missions, the use of cartel drones for intelligence, surveillance, and reconnaissance (ISR) has also been noted.[3] The second trend is the use of *'papas bombas'* (potato bombs) also known as *Papabombas* or *Bombas de impacto* IEDs by organized crime groups, with the Cártel de Jalisco Nueva Generación (CJNG) specifically identified as using them. These IEDs have been utilized by the Fuerzas Armadas Revolucionarias de Colombia (FARC) with their use spreading to CJNG and possibly other groups in the states of Michoacán and Guanajuato in Central Mexico. These IEDs are composed of a sphere-like mass of explosives tightly taped together with the inclusion of nuts and nails for a shrapnel effect. The initial explosive mixture utilized in these devices was potassium chlorate, sulfur, and aluminum powder based. Their use or possession in Mexico has been identified in at least four cases since at least February 2017, two in La Piedad (with one detonation), one in Santiguito, and one near Cerrito de Ortiz, all in Michoacán.[4]

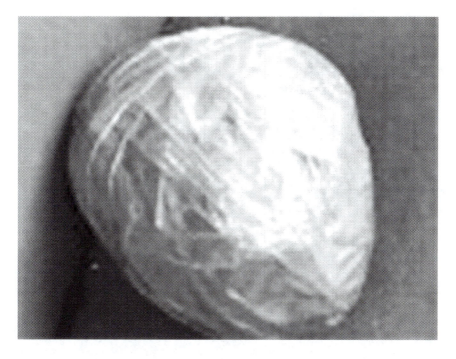

IED ('*Papa Bomba*') Based on FARC Model
Source: Andrés Becerril, "Narcos copian bombas de FARC; Cisen
alerta de explosivos tipo 'papa.'" *Excelsior.* 21 July 2017,
http://www.excelsior.com.mx/nacional/2017/07/21/1176937#imagen-1

Conceptually, commercial drone IED weaponization is by no
means new, and in fact, has become institutionalized by the Islamic
State (IS) in its recent operations in Iraq and Syria. Such use is well
known and has been identified in IS tactical actions taking place as early
as December 2015 in Tishrin, Syria and June 2016 in Khan Touman,
Syria.[5] Such actions and later ones most certainly have contributed in
some way via 'TTP diffusion' to the creation of the weaponized cartel
drone recovered. Specific concerns expressed over such "Mexican cartel
UAV evolutionary potentials" have been noted since August 2014 with
"weapons payload use for small arms and IED attack capabilities"
foreseen as one eventuality.[6]

Islamic State IED Drone in Manbij, Syria
Source: @SDF_Press_1. "Syrian Democratic Forces #SDF
fell down #ISIL Military Drone." *Twitter.* 10 July 2016,
https://twitter.com/ SDF_Press_1/media?lang=en&lang=en [Archived]

An image analysis of the IED drone recovered in Valtierrilla, Guanajuato suggests that it is a 3DR Solo Quadcopter. Such drones which first appeared in about 2015 can be presently purchased for less than $250.00 USD (less than 5,000 MXN) online. This drone has a flight time of about 20 minutes while carrying a payload of up to 420 g (0.926 pounds). It can achieve a top speed of 55 mph (89 km/h) with a range of about .5 miles (.8 km).[7]

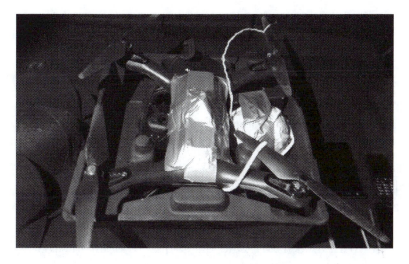

3DR Solo Quadcopter with IED (Front View)
Source: @On_Point_Skillz, "Tál vez a @AbraxasSpa le pueda parecer interesante." *Twitter.* 20 October 2017, https://twitter.com/On_Point_Skillz/status/921521786768056321

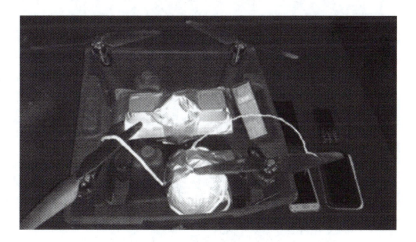

3DR Solo Quadcopter with IED and Remote Detonation Switch (Side View)
Source: "Interceptan ¡dron bomba!"
REDACCIÓN. *Am.* 20 October 2017,
https://www.am.com.mx/2017/10/20/leon/sucesos/interceptan-dron-bomba-385781

The above images of the recovered drone suggest that it is in operational condition. The IED ('*Papa Bomba*') payload is secured to the drone by means of a white rope slung underneath it. No evidence of metal fragmentation components can be seen on the surface of the IED, however, they may have been formed into an outer shell of the IED with additional binding and taping layers built up over them. Underneath the drone a remote control transmitter can be seen although the specific type is not identified—it may be either a 3DR AT10A or AT11A series unit. Two related rectangular components that are next to the drone are not identifiable—possibly they are the controller top piece and a 3DR battery module. Duct tape has been secured to the top of the drone to hold down an unidentified component that is not organic to it—this can be seen with the triangular-like protrusion on the top of the drone under the tape. Speculation exists that this may be the RF detonation receiver for the IED. Yellow electrical wires can also be seen to emanate from the IED. It is further speculated that these yellow wires would be connected to the top RF receiver unit by means of connecting them to RF receiver wires leading down to the underside of the drone. This connection would be made only during the pre-mission arming phase with the yellow wires wound around the white rope securing the IED to the drone. What is assumed to be a detonation control RF emitter unit with 4 toggle switches and a pullout radio antenna can also be seen in one of the images of the recovered IED drone.

Of further note is the imagery of three of the four men arrested in this incident. They all appear physically fit and manifest short military-like haircuts suggesting that they may have either military or law enforcement special operations backgrounds.

Guanajuato is presently plagued with the highest violence levels in the state's history with the following cartels operating in it are "La Familia Michoacana" and/or "Caballeros Templarios," "Cártel del Golfo," "Zetas" and "Cártel Jalisco Nueva Generación (CJNG)."[8] Full scale narco warfare has essentially broken out in this state between competing cartel factions. The latter three cartels are known for actively recruiting cartel enforcers having military and paramilitary backgrounds with CJNG directly linked to recent '*Papas Bombas*' IED-type use.

If the IED drone recovered was ultimately intended for use within the state of Guanajuato, in Morelia, Michoacán, or somewhere else in the state of Michoacán is unknown. What is known is this weaponized drone cost less than $300.00 USD (5,700.00 MXN) to procure—with IED construction costs factored in—and represents a significant escalation in the criminal insurgencies taking place in Mexico. The drone can be flown against its intended target with precision and a high rate of speed and remotely detonated for assassination, anti-personnel, and/or narcoterrorism purposes. It represents another terrorist TTP firebreak crossed in Mexico this year and is indicative of the spiraling levels of narco violence now engulfing wide regions of that sovereign nation.

Significance: Cártel de Jalisco Nueva Generación; CJNG links (Probable), Cartel Technology, Improvised Explosive Device (IED), Papas Bombas, Tactics, Techniques, and Procedures (TTPs), Weaponized Drone, Weaponized Unmanned Aerial System (UAS)

Sources

"Caen cuatro hombres que transportaban un dron acoplado a un artefacto explosivo en Guanajuato." *Proceso.* 20 October 2017, http://www.proceso.com.mx/508237/caen-cuatro-hombres-transport aban-dron-acoplado-a-artefacto-explosivo-en-guanajuato/amp.
"'Dron bomba' listo para detonar a distancia." *AM.* 20 October 2017, https://www.am.com.mx/2017/10/20/sucesos/dron-bomba-listo-para-detonar-a-distancia-385808.
"Dron explosivo: Último artefacto del crimen organizado en México." *HispanTV.* 21 October 2017, http://www.hispantv.com/noticias/mexico/357219/incautan-dron-crimen-organizado-violencia.
"Interceptan ¡dron bomba!" *AM.* 20 October 2017, https://www.am.com.mx/2017/10/20/leon/sucesos/interceptan-dron-bomba-385781.
"¿Tomar video? Dron con fines terroristas." REDACCIÓN. *Am.* 21 October 2017, https://www.am.com.mx/2017/10/21/local/tomar-video-dron-con-fines-terroristas--386171.

"Viajaban en auto robado: les hallan un dron y explosivos." *El Debate*. 21 October 2017, https://www.debate.com.mx/mexico/Viajaban-en-auto-robado-les-hallan-un-dron-y-explosivos-20171021-0007.html.

Endnotes

[1]   Robert J. Bunker and John P. Sullivan, "Mexican Cartel Tactical Note #33: Terrorist TTP Firebreak Crossed - Criminal Group Utilizes Women and Children as Human Shields in Palmarito, Puebla." *Small Wars Journal*. 11 May 2017, http://smallwarsjournal.com/jrnl/art/mexican-cartel-tactical-note-33-terrorist-ttp-firebreak-crossed-criminal-group-utilizes-wom.

[2]   Centro Nacional de Información, "Informe de víctimas de homicidio, secuestro y extorsión 2017." SEGOB/SESNSP. Octubre 2017, http://secretariadoejecutivo.gob.mx/docs/pdfs/victimas/Victimas2017_092017.pdf.

[3]   John P. Sullivan and Robert J. Bunker, "Mexican Cartel Strategic Note No. 18: Narcodrones on the Border and Beyond." *Small Wars Journal*. 28 March 2016, http://smallwarsjournal.com/jrnl/art/mexican-cartel-strategic-note-no-18-narcodrones-on-the-border-and-beyond.

[4]   Andrés Becerril, "Narcos copian bombas de FARC; Cisen alerta de explosivos tipo 'papa.'" *Excelsior*. 21 July 2017, http://www.excelsior.com.mx/nacional/2017/07/21/1176937.

[5]   David Hambling, "ISIS Is Reportedly Packing Drones With Explosives Now." *Popular Mechanics*. 16 December 2015, http://www.popularmechanics.com/military/weapons/a18577/isis-packing-drones-with-explosives/ and @green_lemonnn. "#Syria Sayed Hakim, Fatemiyoun deputy in Palmyra KIA in Palmyra days ago. Reportedly hit by an IS suicide drone." *Twitter*. 9 June 2016, https://twitter.com/warreports/status/740970884798746624.

[6]   Robert J. Bunker, "Mexican Cartel Tactical Note #21: Cartel Unmanned Aerial Vehicles (UAVs)." *Small Wars Journal*. 1 August 2014, http://smallwarsjournal.com/blog/mexican-cartel-tactical-note-21.

[7]   "Solo Specs: Just the facts." *3DR*. 4 May 2015, https://3dr.com/blog/solo-specs-just-the-facts-14480cb55722/#.

[8]   "Guanajuato ranked top ten in violence rate nationwide." *San Miguel Times*. 17 March 2017, http://sanmigueltimes.com/2017/03/guanajuato-ranked-top-ten-in-violence-rate-nationwide/.

Further Reading

Cartel Drones/UAV/UAS

Robert J. Bunker, "Mexican Cartel Tactical Note #21: Cartel Unmanned Aerial Vehicles (UAVs)." *Small Wars Journal.* 1 August 2014, http://smallwarsjournal.com/blog/mexican-cartel-tactical-note-21.

John P. Sullivan and Robert J. Bunker, "Mexican Cartel Strategic Note No. 18: Narcodrones on the Border and Beyond." *Small Wars Journal.* 28 March 2016, http://smallwarsjournal.com/jrnl/art/mexican-cartel-strategic-note-no-18-narcodrones-on-the-border-and-beyond.

Brenda Fiegel, "Narco-Drones: A New Way to Transport Drugs." *Small Wars Journal.* 5 July 2017, http://smallwarsjournal.com/jrnl/art/narco-drones-a-new-way-to-transport-drugs.

Papas Bombas

Santiago Cepeda, "El peligroso mundo de las 'papas bomba'." *Revista DonJuan.* 18 April 2012, http://m.revistadonjuan.com/historias/el-peligroso-mundo-de-las-papas-bomba+articulo+11596882.

Andrés Becerril, "Narcos copian bombas de FARC; Cisen alerta de explosivos tipo 'papa'." *Excelsior.* 21 July 2017, http://www.excelsior.com.mx/nacional/2017/07/21/1176937.

"CJNG utiliza explosivos tipo 'papa' similares a las de las FARC." *El Debate.* 21 July 2017, https://www.debate.com.mx/mexico/CJNG-utiliza-explosivos-tipo-papa-similares-a-las-de-las-FARC-20170721-0066.html.

Chapter 10

Mexican Cartel Tactical Note #38: Armed Drone Targets the Baja California Public Safety Secretary's Residence in Tecate, Mexico

John P. Sullivan, Robert J. Bunker, and David A. Kuhn

Initially Published in Small Wars Journal on 6 August 2018

On Tuesday, 10 July 2018, an armed drone targeted the residence of Gerardo Sosa Olachea, the public safety secretary/*Secretario de Seguridad Pública Estalal* (SSPE) of Baja California, in *colonia* Los Laureles in Tecate—a border city in the San Diego-Tijuana metropolitan area. A second drone, which may have been utilized for ISR (intelligence, surveillance and reconnaissance) and C^2 purposes, was seen over the incident scene. At least one of the drones was equipped with a video camera link and was armed with two IEDs that did not detonate. For a number of international security professionals tracking cartel and gang violence in Mexico—including the authors of this note—an incident like this has been expected for some time now, given the earlier I&W (Indications & Warnings) event that took place in Guanajunto state in October 2017 when a weaponized drone was seized from *Cártel de Jalisco Nueva Generación* (CJNG) operatives.

Key Information: "Con drones envían granadas a casa de Sosa Olachea." *Zeta.* 10 July 2018, http://zetatijuana.com/2018/07/con-drones-envian-granadas-a-casa-de-sosa-olachea/:

> Pasadas las 07:00 horas del martes 10 de julio, dos drones fueron enviados a sobrevolar la casa del Secretario de Seguridad Pública de Baja California, Gerardo Sosa Olachea, ubicada en la colonia Los Laureles, en Tecate, en las inmediaciones de la carretera a Ensenada.
>
> Poco antes de las 07:30 horas cuando testigos pudieron ver cuando los aparatos descendían sobre el inmueble, que en ese momento estaba ocupado por una persona que apoya a la familia con el aseo.
>
> La base de la Policía Estatal Preventiva (PEP) recibió dos reportes al respecto, primero de los artefactos sobrevolando y el segundo indicando que en la casa de Sosa Olachea habían localizado por lo menos uno de los drones equipados con cámara y audio. En este aparato, sujetaron con cinta adhesiva dos granadas de fragmentación que no detonaron cuando los dejaron caer.

Key Information: "Reportan drones con granadas; uno descendió en la casa del titular de la SSP en BC." *Proceso.* 10 July 2018, https://www.proceso.com.mx/542389/reportan-drones-con-granadas-uno-descendio-en-la-casa-del-titular-de-la-ssp-en-bc:

> Los drones serán revisados por la Policía Cibernética de la SSP para conocer su origen y la ruta que siguieron hasta llegar al espacio aéreo de la casa de Sosa Olachea.
>
> Pese a que los drones son de capacidad media y requieren que la persona que los maneja esté cerca del perímetro de vuelo, nadie fue aprehendido. Al parecer

los agresores huyeron al momento de observar que las granadas no detonaron…

Sosa Olachea dijo al semanario Zeta que no ha recibido amenazas desde que tomó el cargo, en octubre de 2017, pero presume que el ataque pudo venir de los cárteles de las drogas "a los que les hemos pegado mucho con los decomisos, que más que grandes son constantes, del narcomenudeo, de los laboratorios localizados en Tecate y Rosario".

Key Information: "Con Granadas Transportadas En Un Dron Atacan Casa De Secretario De Seguridad Pública De BC." *Reporte Indigo*. 10 July 2018, https://www.reporteindigo.com/reporte/granadas-transportadas-en-dron-atacan-casa-secretario-seguridad-publica-bc/:

La madrugada de este martes **dos granadas** fueron colocadas en el interior de una de las casas del secretario de Seguridad Pública de Baja California, **Gerardo Manuel Sosa Olachea**.

Para lograr su cometido, los artífices del ataque utilizaron un **dron** en el que amarraron los artefactos explosivos para dejarlos caer en la propiedad, ubicada en el municipio de Tecate.

Key Information: Ángel F. González, "Señala SSPE a cárteles en atentado a secretario." *Frontera.info*. 11 July 2018, http://www.frontera.info/EdicionEnLinea/Notas/Noticias/11072018/1355958-Senala-SSPE-a-carteles-en-atentado-a-secretario.html:

La vulnerabilidad

El hecho de que integrantes del crimen organizado dejaran un dispositivo con granadas de fragmentación en la casa del titular de la Secretaría de Seguridad Pública Estatal (SSPE) demuestra la vulnerabilidad en la que se encuentra la ciudadanía en Baja California, consideró el presidente del Consejo Ciudadano de Seguridad Pública, Juan Manuel Hernández Niebla.

"Cuando el crimen organizado pretende amenazar a un alto dirigente policiaco, refleja el grado de vulnerabilidad que tenemos los ciudadanos en Baja California", dijo.

Tras el hallazgo del dron, añadió, es necesario que se descubra a los presuntos responsables para que las bandas del crimen organizado tengan en claro que esas acciones no quedarán impunes.

Key Information: "La amenaza." *Zeta*, 16 July 2018, http://zetatijuana. com/2018/07/la-amenaza-3/.

El 10 de julio de 2018, sin más preámbulo y utilizando un dron, alguien sobrevoló el espacio aéreo de la residencia de Sosa en Tecate, e hizo descender el artefacto ahí mismo. Pegadas iban las que parecían ser dos granadas de fragmentación. Aunque inicialmente lo consideró un atentado, el caso terminó en una amenaza. En el dron había un mensaje con imágenes de cercanos al funcionario estatal…

Investigadores reiteraron que en Tecate solo existe una célula criminal "en control, pero se han pasado de un cártel a otro según les convenga, actualmente están con el Cártel Jalisco."

Who: The attack was conducted by a criminal organization—a Mexican cartel—with the ability and willingness to engage in a tactical action in the contested Tecate segment of the Tijuana *plaza* against the senior ranking state law enforcement official. The dominant cartels operating in Baja California state are *Cártel de Sinaloa* (CDS) and remnants of the Tijuana Cartel/*Cártel Arellano Félix* (CAF) or Arellano-Félix Organization (AFO) which are in the process of reforming under the banner of the *Cártel de Jalisco Nueva Generación* (CJNG) as the *Cártel de Tijuana Nueva Generación* (CTNG). Considering past patterns of violence in the Tecate area and the broader Tijuana *plaza*—as well as in Baja California itself—this attack has a signature consistent with prior CTNG/CJNG activity and can likely be attributed to a CTNG/CJNG operational cell.

What: A weaponized drone attack targeting the residence (known as "La Quinta Marina") of the public safety secretary of Baja California (*Secretario de Seguridad Pública Estalal*–SSPE) Gerardo Sosa Olachea. This is the first recorded incident in Mexico in which a weaponized cartel drone has both been used a) offensively and b) in a coordinated manner with a second mission support drone.

When: The attack took place at 0700-0730 on Tuesday 10 July 2018. The attack was timed while the Baja California public safety secretary was leaving his residence and getting into his vehicle in order to later meet with the new US Consul in Tijuana. According to the Editor-in-Chief of *Zeta* (the Tijuana news magazine), the timing of attack was closer to 0740.[1] The fact the public safety secretary was even at home at this time is in direct variance to an official Mexican governmental release that said the secretary what not a home at the time of the incident and actually lives in Mexicali (some 68.7 miles away).[2]

Where: The residence of the *Secretario de Seguridad Pública Estalal* (SSPE) of Baja California is located in Tecate, Mexico in the *colonia* Los Laureles subdivision and is well known to the local residents. This border city is 23.7 miles east of Tijuana in the San Diego-Tijuana metropolitan area and has a present population of approximately 75,000.

Why: Based on an ongoing pattern of earlier threats directed at lower level police officials by organized crime—specifically the CTNG/CJNG alliance—and the inoperability (render safe procedures) of the IEDs carried on the attacking drone, this incident can be viewed as a threat, a communication or 'threat escalation' meant to intimidate the public safety secretary of Baja California rather than an actual attack (or attack dry run) in which mission failure took place.

Analysis (Background)

Drones are increasingly a feature of the narco-conflict in the *plazas* (the narco-market controlled by a single cartel or gangs or contested among many seeking dominance) along the US-Mexico border.[3] As recently as late June, the *Washington Post* reported that illicit drone flights were surging.[4] Most narcodrones are used for intelligence, surveillance, and reconnaissance (ISR) or for light trafficking of contraband or drugs, however the potential for armed drone use by cartels and gangs has existed for several years. That potential was strongly confirmed in 20 October 2017 when Mexican Federal Police/ *Policía Federal* (PF) in Guanajunto interdicted an improvised explosive drone—'*dron bomba*'—consistent with potato bombs (*papas bombas*) used by the *Cártel de Jalisco Nueva Generación* (CJNG).[5]

Tecate, like much of the Tijuana plaza, is contested turf. The Sinaloa cartel, which gained dominance of the *plaza* after battling the Arellano-Félix Organization (AFO)/*Cártel Arellano Félix* (CAF) or Tijuana Cartel), is once again being challenged by remnants of the AFO known as the *Cártel de Tijuana Nueva Generación* (New Generation Tijuana Cartel) under the protection of the CJNG.[6] Indeed, the police have long been targeted in the narco-violence in this border city just south of San Diego. In addition to the drone attack on the home of SSPE secretary Sosa Olachea on Tuesday, 10 July 2018, a state ministerial police officer was injured in an armed attack at about 1800 hours on 29 June 2018. Witnesses said two suspects fired on the commander of the *Policía Ministerial* who repelled the attack while a uniformed officer was injured and transported to hospital.[7] In June 2017, a *sicario* working for the Cártel de Sinaloa was captured, tortured, sexually assaulted, and beheaded by suspected CJNG operatives. His head was left near the home of Tecate's municipal public safety director along with a *narcomensaje* (narco message). The gruesome display was interpreted as a warning to the police director.[8]

Analysis (Technical Assessment)

Two drones targeted the home of SSPE secretary Gerardo Sosa Olachea in the Tecate neighborhood of *colonia* Los Laureles, in Tecate at approximately 0700-0730 hours (though this incident may have taken place as late as 0740). One of the drones landed on the property—known as "La Quinta Marina"—as the secretary was leaving for a meeting with the new Consul of the United States in Tijuana. According to multiple press reports, the devices on one of the drones failed to detonate and the area was cordoned by state preventive police/ *Policía Estatal Preventiva*(PEP). At the time of this report, one of the drones and the devices it held will be undergoing forensic assessment by SEDENA (*Secretaría de la Defensa Nacional*). The cyber squad of the SSPE (*Policía Cibernética*) is also expected to examine the drone.[9] Images of this drone have been graciously provided courtesy of César R. Blanco Villalón, Editor-in-Chief, *Zeta*, for this tactical note.[10]

Attacking Tarot Ironman Drone Hard Landing in
a Courtyard with Tennis Court in the Background;
Carbon Fiber Landing Gear Leg Broken Off
Photograph 1: Courtesy of *Zeta*

Close Up of Attacking Tarot Ironman Drone; *Tactical Note Forensic Notations (Green Blanket—Post-Crash Placement—Probable Video Camera Feed Obscurement; 1ˢᵗ IED taped on the left-side; 2ⁿᵈ IED taped on the right-side; White Tube also taped to drone)*
Photograph 2: Courtesy of *Zeta*

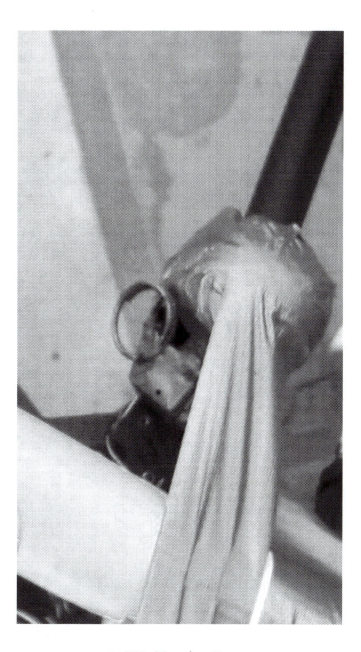

1st IED Taped to Frame
Photograph 3: Courtesy of *Zeta*

2nd IED Taped to the Frame
Photograph 4: Courtesy of *Zeta*

Various drone technical and incident TTP (tactics, techniques, and procedures) specifics have been determined concerning this incident. The drone in question made either a hard landing or was purposely put down hard by command signal in order to prevent any possibility of further movement (as the drone's onboard radio link is still active) in the courtyard of the residence. Forensic imagery analysis of the drone provided by Brandon "Sas" Sasnett, Director of Threat Analysis, Red Six Solutions, LLC yields the following information:

Airframe: Tarot Ironman
Video System: DJI AVL58
Flight Controller: DJI Naza-M
Receiver: FrSky V8FR
Transmitter: Taranis (Compatible)
Maximum Flight Distance: Tecate Environment; Appx. 1.3 KM (0.8 miles) [11]

The drone—a hexacopter (i.e. configured with six rotors)—is kit based with the airframe and components manufactured by Tarot (Wenzhou, Zhejiang province, China; www.tarot-rc.com) but can be configured with other radio controlled hobbyist products. It carries a typical payload of 1.5 to 2.5kg (3.3 to 5.5 lbs) for the 680 pro series— and increasingly more for the later 810/960/1000 series. The drone may be of the X6 or 690 type but this is a speculative assumption only. The Tarot Ironman has a general flight speed of about 20-25 MPH, with the basic airframe retailing in the low-to-mid hundreds of dollars and the basic platform assembled to meet user needs. A fully configured drone, along with the flight controller system, can range anywhere from the high-hundreds of dollars into the low thousands, depending on whether it is built from scratch from parts or purchased as a completed unit. The landing gear of the drone was in the down position when it hard landed.

An analysis of the IEDs placed on the drone confirms that they were attached to two of the six frame arms and that a total of two different types of fragmentation hand grenades were utilized. The *baseball* style hand grenade is an M-67 fragmentation grenade (or a foreign copy thereof), and the second is a Chilean MK-2 fragmentation grenade. The M-67 (or similar copies) is relatively easy to obtain throughout Central and South America. The Chilean MK-2 is a close copy of the WWII and Korean era U.S. MK-2 grenades. This version was still in production and available as of 2002 from the manufacturer— Cardoen Industries (*Industrias Cardoen*), a South American defense company located in Piso, Santiago. It is important to note that this munition, although somewhat dated, is considered highly reliable if stored with minimal care.

The Chilean MK-2 grenade has several distinctive features that are not common to other MK-2 copies. The safety lever, for example, follows a European style that captures the top of the fuze housing by hooking over the top, as opposed to using a fork that hooks under a T-bar. Additionally, the fuze housing of the Chilean version flares to a lip at the base that protrudes past the truncated top of the grenade body. Both of these features can be observed in the highlighted areas in enhanced Photographs: 4 and 4A.

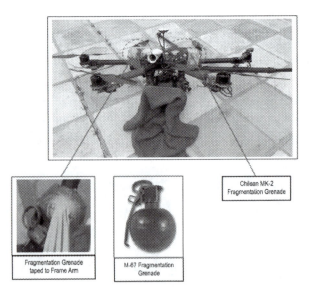

Additional Photograph 2 (& 3) Analysis with M-67 Fragmentation Grenade Identification (Original Photos Courtesy of *Zeta*)

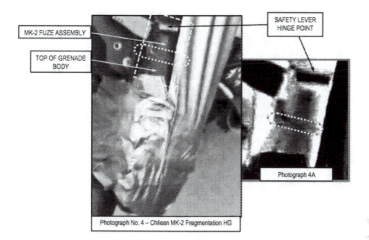

*Additional Photograph 4 Analysis with 4A Imaging
for MK-2 Grenade Identification*
(Original Photo Courtesy of *Zeta*)

The two grenades that are attached to the *La Quinta Marina* drone have been secured to the frame arms with the safety pins in place. Based upon the forceful landing combined with the intact safety pins, it appears in all respects that the operators of the drone wanted to send a clear message to the resident that a fully armed drone of similar configuration could be sent at any time. It is also believed that the grenades aboard this drone are live, as *inert* ones would certainly not carry the same message impact.

The reader should note that, although the safety levers of both hand grenades were taped over in the process of securing the grenades to the drone, there are a number of methods by which the grenades (armed with pins removed) could be detonated either at low altitude or following a landing in a specific location. Due to reasons of security, these will not be addressed here. There are also drone configurations previously used by ISIS that allow explosive payloads to be released at altitude against a specific target.

Typically, two drones are used in any of the above attack methods. The second drone is used not only for spotting and documenting, but to assist the primary drone operator in establishing and maintaining correct flight orientation. The reader will note that the drone in the photograph has two of the six beam arms marked in green while the opposing two arms display red markings that allow the operator to verify forward and aft orientation for positive visual flight control from a distance.

The M-67 fragmentation grenade contains a filler of 184.6 gm. of Composition "B" explosive and has a casualty radius of 15-meters. The Chilean Mk-2 contains a 115gm. explosive filler that is believed to be either flaked TNT or Composition "B." This grenade has a similar casualty radius, although not as efficient in shrapnel dispersal as the M-67.

The total combined weight of the two grenades is 1,052 grams (2.32 lbs.). This is well under the 5 kg. (11 lb.) payload capability of later 690S or X6 series and still within the normal payload parameters of the earlier 680 pro series.[12]

The white tube—composed either of rolled cardboard or plastic—which was taped to the top of the drone (refer to Photographs 1 & 2) was said to supposedly contain a narco message (*narcomensaje*) and pictures of Gerardo Sosa Olachea's family members taken from their private Facebook accounts.[13] This information, as well as the physical fact that the IEDs were rendered inoperable, leads credence to the 'threat escalation' hypothesis endorsed in this tactical note.

The operational cell engaged in this mission was likely composed of at least five individuals—two drone pilots, a driver, and two dedicated security—and one transport vehicle such as an SUV, though for coordination purposes ground surveillance of the targeted primary residence would be utilized as well as neighborhood lookouts (*halcones*) deployed to provide early warning of approaching police patrols. Given the limitations of the drones utilized—which possesses roughly 20 minute flight times—an operational window of only about 10-15 minutes existed, as launch and recovery windows would also have to be taken into consideration. This would necessitate a launch point within a quarter to half a mile (0.25-0.50 miles) of the residence rather than at maximum flight distances (0.8 miles) to better ensure operational success. Further characterization of this suspected operational cell, once again due to security considerations, will not be provided.

Conclusion

While the intent of the attackers is unknown at this time (SEDENA, the SSPE, and PEP have not disclosed the identity of the suspected criminal organization pending further investigation), it is highly probable that the attack was a demonstration of intent or a warning to the SSPE secretary as seen in the CJNG warning to the Tecate police director in late June 2017.[14] Potential actors behind the attack/ threat include 1) state police working for narcos; 2) local criminals from Tecate; 3) *huachicoleros* (fuel thieves); 4) *narcomenudistas* (street drug dealers); and 5) narco cartels (*cárteles de la droga*) including the CJNG, *Cártel de Sinaloa* (CDS), or *Cártel Arellano Félix* (CAF). The Tecate area is currently assessed to be dominated by the CJNG, that is, local criminal organizations, such as the *Los Enfermos, Los Hemanos Valdez*, and *La Bondad* currently owe allegiance to *capos* under the CJNG umbrella.[15] This territorial control combined with prior CJNG targeting and drone potentials makes the CJNG the likely author of this attack/threat demonstration.

While it is possible that the device(s) failed and the attackers intended direct harm or even that a dry run was being engaged in with the weaponized drone simulating an attack crashing to the ground, the IED safety protocols utilized and threatening note to Gerardo Sosa Olachea, accompanied by photos of his family, definitely suggest otherwise. This attack is significant in the direct targeting of a state police official by a criminal organization in a contested *plaza*.

This incident should be considered an escalation of cartel/gang drone use and certainly won't be the last use of armed drones in Mexico's crime wars or by terrorists and/or insurgents elsewhere.[16] Indeed, as we complete this assessment, a drone attack on 4 August 2018 in Caracas characterized as an attempted assassination on Venezuelan President Nicolas Maduro dominates the news.[17] The effectiveness of these future drone attacks is expected to vary—indeed most can be expected to inflict limited damage—however over time the threat will likely mature, yielding enhanced lethality and operational effectiveness.[18] Further, from a homeland security perspective, this incident represents the first confirmed offensive use of a weaponized cartel drone in Mexico—one taking place only a few miles below the US-Mexico border.

Significance: Cártel de Jalisco Nueva Generación; CJNG Links (*Potentials*), Cartel Technology, Drones, Improvised Explosive Devices (IEDs), Narcodrones, Tactics, Techniques, and Procedures (TTPs), Weaponized Drones, Weaponized Unmanned Aerial System (UAS)

Incident overview, I&W, and past narco drone use analysis conducted by John P. Sullivan and Robert J. Bunker. Drone IED forensics and analysis conducted by David A. Kuhn.

Sources

"La amenaza." *Zeta*. 16 July 2018, http://zetatijuana.com/2018/07/la-amenaza-3/.

Ángel F. González, "Señala SSPE a cárteles en atentado a secretario." *Frontera.info*. 11 July 2018, http://www.frontera.info/EdicionEnLinea/Notas/Noticias/11072018/1355958-Senala-SSPE-a-carteles-en-atentado-a-secretario.html.

"Con drones envían granadas a casa de Sosa Olachea." *Zeta*. 10 July 2018, http://zetatijuana.com/2018/07/con-drones-envian-granadas-a-casa-de-sosa-olachea/.

"Con Granadas Transportadas En Un Dron Atacan Casa De Secretario De Seguridad Pública De BC." *Reporte Indigo*. 10 July 2018, https://www.reporteindigo.com/reporte/granadas-transportadas-en-dron-atacan-casa-secretario-seguridad-publica-bc/.

"Desconocidos pegan granadas a dron y las hacen caer en casa del Secretario de Seguridad Pública de BC." *Sin Embargo*. 10 July 2018, http://www.sinembargo.mx/10-07-2018/3440637.

José Jiménez Díaz, "No detona dron con granadas en casa de Sosa Olachea." *El Mexicano*. 10 July 2018, http://www.el-mexicano.com.mx/informacion/noticias/1/3/estatal/2018/07/10/1052433/no-detona-dron-con-granadas-en-casa-de-sosa-olachea.

"Reportan drones con granadas; uno descendió en la casa del titular de la SSP en BC." *Proceso.* 10 July 2018, https://www.proceso.com. mx/542389/reportan-drones-con-granadas-uno-descendio-en-la-casa-del-titular-de-la-ssp-en-bc.

Endnotes

[1] Robert J. Bunker email correspondence with César R. Blanco Villalón, Editor-in-Chief, *Zeta*, on 13 July 2018. The new Council of the United States in Tijuana is Sue Saarnio. See https://mx.usembassy.gov/embassy-consulates/tijuana/consul-general-sue-saarnio/.

[2] "Drone with grenades falls into official's residence." *Imperial Valley Press.* 11 July 2018, https://www.pressreader.com/usa/imperial-valley-press/20180711/281582356394974.

[3] See Robert J. Bunker, "Mexican Cartel Tactical Note #21: Cartel Unmanned Aerial Vehicles (UAVs)." *Small Wars Journal.* 1 August 2014, http://smallwarsjournal.com/blog/mexican-cartel-tactical-note-21; John P. Sullivan and Robert J. Bunker, "Mexican Cartel Strategic Note No. 18: Narcodrones on the Border and Beyond." *Small Wars Journal.* 28 March 2016, http://smallwarsjournal.com/jrnl/art/mexican-cartel-strategic-note-no-18-narcodrones-on-the-border-and-beyond; and Aaron R. Schmershi, *Fifty feet above the wall: cartel drones in the U.S.-Mexico border zone airspace, and what to do about them.* Masters Thesis. Monterey: Naval Postgraduate School, 2018, http://hdl.handle.net/10945/58364.

[4] Gina Harkins, "Illicit drone flights surge along U.S.-Mexico border as smugglers hunt for soft spots." *Washington Post.* 24 June 2018, https://www.washingtonpost.com/world/national-security/illicit-drone-flights-surge-along-us-mexico-border-as-smugglers-hunt-for-soft-spots/2018/06/24/ea353d2a-70aa-11e8-bd50-b80389a4e569_story.html?utm_term=.0992752ea810.

[5] Robert J. Bunker and John P. Sullivan, "Mexican Cartel Tactical Note #35: Weaponized Drone/UAV/UAS Seized in Valtierrilla, Guanajuato with Remote Detonation IED ('Papa Bomba') Payload." *Small Wars Journal.* 23 October 2017, http://smallwarsjournal.com/jrnl/art/mexican-cartel-tactical-note-35.

[6] "El Cártel de Sinaloa y el Cártel Jalisco Nueva Generación se pelean por la plaza de Tecate" in "El narco amenaza al gobierno en Tecate, Baja California." *El Debate.* 28 September 2016, https://www.debate.com.mx/policiacas/El-narco-amenaza-al-gobierno-en-Tecate-Baja-California-20160928-0223.html; "Tijuana, territorio en disputa por el CJNG y el Cártel de Sinaloa." *El Financiero.* 25 February 2018, http://www.elfinanciero.com.mx/nacional/el-cartel-de-jalisco-nueva-generacion-intenta-arrebatarle-tijuana-al-cartel-de-sinaloa; and

"Recrudece violencia Cártel Jalisco." *Zeta*, 18 June 2018, http://zetatijuana. com/2018/06/recrudece-violencia-cartel-jalisco/.

[7] "Atentan contra Comandante de la Ministerial en Tecate; huyen a Tijuana." *Uniradio Informa*, 29 June 2018, http://www.uniradioinforma.com/noticias/ policiaca/530043/atentan-contra-comandante-de-la-ministerial-en-tecate-huyen-a-tijuana.html and Kathia Borbolla, "Balean a policía ministerial en Tecate." *El Debate*, 20 June 2018, https://www.debate.com.mx/policiacas/policia-ministerial-baleado-ataque-armado-comandante-tecate--20180630-0059.html.

[8] Carlos Lima and Esther Hernández. "Abandonan cabeza cerca de casa del jefe de Policía de Tecate." *Frontera.info*. 6 October 2017, http://www.frontera. info/EdicionEnLinea/Notas/Noticias/10062017/1224350-Abandonan-cabeza-cerca-de-casa-del-jefe-de-Policia-de-Tecate.html and "Cae cabeza del CJNG en Tecate." *Zeta*. 26 June 2017, http://zetatijuana.com/2017/06/ cae-cabeza-del-cjng-en-tecate/.

[9] See the Key Information entries and sources listed in the text for specific references.

[10] César R. Blanco Villalón, Editor-in-Chief, *Zeta*, email correspondence with Robert J. Bunker on 13 July 2018.

[11] Brandon "Sas" Sasnett, Director of Threat Analysis, Red Six Solutions, LLC email correspondence with Robert J. Bunker on 10 July 2018.

[12] It should be noted that the continued advent of higher technology drones entering the world market that are capable of higher payloads, longer flight times, and increased guidance capability will prove a continuing and significant threat when placed in use by terrorists and now narco-terrorists and criminal insurgents.

[13] César R. Blanco Villalón, Editor-in-Chief, *Zeta*, email correspondence with Robert J. Bunker on 13 July 2018. Additionally, in this correspondence it was independently confirmed that, "The grenades were manipulated, they didn't have the capacity to blow up."

[14] Op. cit., Note 8.

[15] "Investigadores reiteraron que en Tecate solo existe una célula criminal "en control, pero se han pasado de un cártel a otro según les convenga, actualmente están con el Cártel Jalisco" at "La amenaza." *Zeta*, 16 July 2018, http:// zetatijuana.com/2018/07/la-amenaza-3/.

[16] See Colin P. Clarke, "A Terrorist's Dream: How Twitter and Toy Drones Could Kill a Lot of People." *The National Interest*. 17 June 2018, https:// nationalinterest.org/blog/the-buzz/terrorists-dream-how-twitter-toy-drones-could-kill-lot-26288?nopaging=1.

[17] Alexandra Ulmer and Vivian Sequera, "Venezuela's Maduro says drone blast was bid to kill him, blames Colombia," Reuters Venezuela. 4 August 2018, https://www.reuters.com/article/us-venezuela-politics/

venezuelas-maduro-target-of-drone-attack-but-unharmed-government-idUSKBN1KP0SA; Scott Smith and Christine Armario, "Venezuela's Maduro: Drone attack was attempt to kill him." *Washington Post*. 4 August 2018, https://www.washingtonpost.com/world/the_americas/venezuelan-government-drone-strikes-targeted-maduro/2018/08/04/01034d9a-9846-11e8-818b-e9b7348cd87d_story.html?utm_term=.b55612dac292; and Ana Vanessa Herrero and Nicholas Casey, "Venezuelan President Targeted by Drone Attack, Officials Say." *New York Times*. 4 August 2018, https://www.nytimes.com/2018/08/04/world/americas/venezuelan-president-targeted-in-attack-attempt-minister-says.html.

[18] See Scott Stewart, "When Drones Attack: The Threat Remains Limited." *Stratfor Worldview*. 17 July 2018, https://worldview.stratfor.com/article/when-drones-attack-threat-remains-limited.

For Additional Reading

Robert J. Bunker and John P. Sullivan, "Mexican Cartel Tactical Note #35: Weaponized Drone/UAV/UAS Seized in Valtierrilla, Guanajuato with Remote Detonation IED ('Papa Bomba') Payload." *Small Wars Journal*. 23 October 2017, http://smallwarsjournal.com/jrnl/art/mexican-cartel-tactical-note-35.

John P. Sullivan and Robert J. Bunker, "Mexican Cartel Strategic Note No. 18: Narcodrones on the Border and Beyond." *Small Wars Journal*. 28 March 2016, http://smallwarsjournal.com/jrnl/art/mexican-cartel-strategic-note-no-18-narcodrones-on-the-border-and-beyond.

Robert J. Bunker, *Terrorist and Insurgent Unmanned Aerial Vehicles: Use, Potentials, and Military Implications*. Carlisle Barracks: Strategic Studies Institute, US Army War College. 27 August 2015, https://ssi.armywarcollege.edu/pubs/display.cfm?pubID=1287.

Robert J. Bunker, "Mexican Cartel Tactical Note #21: Cartel Unmanned Aerial Vehicles (UAVs)." *Small Wars Journal*. 1 August 2014, http://smallwarsjournal.com/blog/mexican-cartel-tactical-note-21.

Chapter 11

Are Armed Drones the Weapon of the Future for Mexico's Cartels?

Parker Asmann

Initially Published in InSight Crime on 15 August 2018

Are drones the weapon of the future for Mexico's criminal groups?

Mexico's powerful drug cartels could be using armed drones to attack those impeding their criminal operations, marking the potential expansion of the use of this technology from just transporting drugs or carrying out surveillance.

In July of this year, an armed drone was used to attack the house of Baja California state Public Safety Secretary Gerardo Sosa Olachea in the city of Tecate along the US-Mexico border, *Zeta* magazine[1] reported.

At least two drones were allegedly used in the strike. The first was armed with audio and video equipment and two improvised explosive devices (IEDs) that did not explode after falling into the official's yard. A second drone was seen above the house likely performing surveillance, according to Zeta.

The July incident was the first time that a Mexican cartel used a weaponized drone to carry out such an attack, according to Mike Vigil, the former Chief of International Operations at the US Drug Enforcement Administration (DEA). However, the "inoperability" of the IEDs suggests that the attack was more likely a method to intimidate the official rather than to carry out an actual attack, according to *Small Wars Journal*.[2]

Sosa Olachea told Zeta that he thought the threat likely came from one of the drug cartels operating in the area that officials have dealt serious blows against through seizing drug shipments and dismantling drug laboratories.

The Tijuana Cartel, Sinaloa Cartel and Jalisco Cartel New Generation (Cartel Jalisco Nueva Generación – CJNG) are all known to be operating in this specific area along the US-Mexico border.

Towards the end of 2017, the Daily Beast reported[3] on the first apparent incident in which an armed drone intended for use in a violent attack was discovered in the arsenal of a Mexican cartel.

InSight Crime Analysis

Drones are gaining popularity[4] throughout Latin America among criminal groups and law enforcement officials for their ability to traffic drugs and carry out surveillance. But their use, or intended use, by organized crime to execute violent attacks is a recent development.

A DEA source explained that since 2014, Mexico's cartels have hired local workers to construct custom-made narco drones[5] to traffic cocaine and other drugs across the US-Mexico border.

Mexico's security forces have also relied on[6] drones that transmit images and videos in real-time in order to identify key locations used by the cartels to store drugs and weapons or to carry out other criminal activities. US authorities have also utilized drones[7] in Mexico to aid in the so-called "war on drugs" against the country's criminal groups.

But the latest evidence suggests that drones might be the "way of the future" for Mexico's cartels for engaging in "cartel warfare," according to Vigil.

As Mexico's criminal world continues to fragment, Vigil told *InSight Crime* that these groups may start to use drones as an "offensive weapon simply because it will not require the loss of human personnel, which is difficult to replace especially if they are trained and have a full understanding of the cartel's operations."

Endnotes

[1] "Con drones envían granadas a casa de Sosa Olachea." *Zeta*. 10 July 2018, https://zetatijuana.com/2018/07/con-drones-envian-granadas-a-casa-de-sosa-olachea/.

[2] John P. Sullivan, Robert J. Bunker and David A. Kuhn, "Mexican Cartel Tactical Note #38: Armed Drone Targets the Baja California Public Safety Secretary's Residence in Tecate, Mexico." *Small Wars Journal*. 6 August 2018, https://smallwarsjournal.com/jrnl/art/mexican-cartel-tactical-note-38-armed-drone-targets-baja-california-public-safety.

[3] Jeremy Kryt, "Game of Drones: Mexico's Cartels Have a Deadly New Weapon." *The Daily Beast*. 12 November 2017, https://www.thedailybeast.com/game-of-drones-mexicos-cartels-have-a-deadly-new-weapon.

[4] Marguerite Cawley, "Drone Use in Latin America: Dangers and Opportunities." *InSight Crime*. 18 April 2014, https://www.insightcrime.org/news/analysis/drone-use-in-latin-america-dangers-and-opportunities/.

[5] Camilo Mejia Giraldo, "Mexico's Cartels Building Custom-Made Narco Drones: DEA." *InSight Crime*. 11 July 2014, https://www.insightcrime.org/news/brief/mexico-s-cartels-building-custom-made-narco-drones-dea/.

[6] Jeanna Cullinan, "Mexico to Use Drones Against Drug Cartels." *InSight Crime*. 17 November 2011, https://www.insightcrime.org/news/brief/mexico-to-use-drones-against-drug-cartels/.

[7] "US Drones in Mexico Spark Debate." *InSight Crime*. 17 March 2011, https://www.insightcrime.org/news/analysis/us-drones-in-mexico-spark-debate/.

Chapter 12

El Paso Sector Border Patrol Encounters New Tactics as Smugglers Keep Sending in Families and Felons

U.S. Customs and Border Protection

Initially Published in CBP Newsroom on 17 April 2019 (Release Date)[1]

EL PASO, Texas – Agents have discovered a new tactic in counter surveillance as they continue to apprehend large groups and gang members attempting to illegally enter the United States.

The object traveled approximately 100 yards over
U.S. soil and then return back to Mexico.

As one agent was monitoring the border at night utilizing an infrared camera, he observed a small airborne object traveling northbound across the U.S. International boundary. The object traveled approximately 100 yards over U.S. soil and then return back to Mexico. The object repeated this pattern three times. Approximately two minutes after the object returned back to Mexico for the third time, a group of 10 subjects made an illegal entry into the U.S. in the same area in which the object had been traveling. All 10 subjects were subsequently taken into custody by Border Patrol Agents. This is the first known time in recent history that a drone has been utilized as a 'look-out' in order to aid in illegal entries in the El Paso Sector.

Tuesday began with two large groups shortly before 1 a.m. The first group of over 230 people was encountered at the Antelope Wells Port of Entry in the boot heel of New Mexico at 12:45 a.m. The second was encountered just west of Mount Cristo Rey in Sunland Park, New Mexico, consisting of over 360 people around 12:50 am. The Antelope Wells Port of Entry would encounter a second large group of over 130 subjects right before the end of day at 11:45 p.m. The El Paso Sector had over 1,800 apprehensions in total for the entire day of Tuesday, over half of those being in the El Paso Metropolitan area...

Endnotes

[1]   Last modified: 18 April 2019, https://www.cbp.gov/newsroom/local-media-release/el-paso-sector-border-patrol-encounters-new-tactics-smugglers-keep?_ga=2.256713677.718726486.1600875406-1805667773.1599066724.

Chapter 13

Tuxpan and Tecalitlán, Jalisco: "Thank you, Señor Mencho."

Chivis Martinez

Initially Published in Borderland Beat on 3 May 2020

Gracias Señor Mencho

Using drones, CJNG created videos featuring recipients of the cartels Covid-19 food packages. CJNG has been the most active cartels giving out these "pantries", the event in the videos were in Tecalitlán Jalisco targeting people living in poverty, the same as others dispensed.

As all other propaganda events, the narcos conduct the distributions in complete impunity, without the arrival of police or military elements.

The first video depicts the narcos convoy rolling into town children chasing the vehicles, and ending with people saying "Thank you Mr. Mencho".

The second consists of various people saying thank you for the help.

La Voz Del Pueblo (Oficial)
@L.Pueblo2

#OJO #Tecalitlán #Jalisco

Como si fuera un spot con drones graban vídeo
"Gracias Señor Mencho".

Otra vez y con total impunidad el 29 Abril 2020 integrantes
del Cártel Jalisco Nueva Generación #CJNG repartieron
despensas a familias de escasos recursos por el
#COVID19mx

Vídeo 1

10:37 PM · May 1, 2020

♡ 72   ♀ 32 people are Tweeting about this

La Voz Del Pueblo (Oficial)
@L.Pueblo2

#OJO #Tecalitlán #Jalisco

Como si fuera un spot con drones graban vídeo
"Gracias Señor Mencho".

Otra vez y con total impunidad el 29 Abril 2020 integrantes
del Cártel Jalisco Nueva Generación #CJNG repartieron
despensas a familias de escasos recursos por el
#COVID19mx

Vídeo 2

10:44 PM · May 1, 2020

♡ 71   ♀ 31 people are Tweeting about this

Addendum (Anthology)

Cartel drone video imagery:

Vídeo 1

10:37 PM · May 1, 2020

Vídeo 1

10:37 PM · May 1, 2020

Vídeo 2

6.3K views   0:23 / 2:19

10:44 PM · May 1, 2020

Vídeo 2

6.3K views   2:04 / 2:19

10:44 PM · May 1, 2020

Addendum (Anthology) Translation

Thank you Mister Mencho

Image 1

 The text reads: "As if it were a spot with drones they record video. Thank you Mr. Mencho." [...] "Again and with total impunity On 29 April 2020 members of the Cártel Jalisco New Generation #CJNG distributed food to low-income families for the #COVID19mx."[...] "Video 1." Source: La Voz Del Pueblo, "Como si fuera un spot con drones graban video 'Gracias Señor Mencho.'" *Twitter.* 1 May 2020, https://twitter.com/LPueblo2/status/1256457862630510592?s=20.

Image 2

 The text reads: "As if it were a spot with drones they record video. Thank you Mr. Mencho." [...] "Again and with total impunity On 29 April 2020 members of the Cártel Jalisco New Generation #CJNG distributed food to low-income families for the #COVID19mx."[...] "Video 2." Source: La Voz Del Pueblo, "Como si fuera un spot con drones graban video 'Gracias Señor Mencho.'" *Twitter.* 1 May 2020, https://twitter.com/LPueblo2/status/1256459489856950272?s=20.

Chapter 14

Yuma Sector Agents Intercept Narcotics Dropped From Drones

U.S. Customs and Border Protection

Initially Published in CBP Newsroom on 4 May 2020 (Release Date)[1]

YUMA, Ariz. – Yuma Sector Border Patrol agents working various shifts intercepted narcotics in two of three separate incidents of Single Unmanned Aircraft Systems (SUAS) detected between April 29th and May 3rd. The SUAS are commonly referred to as, 'Drones'.

On Wednesday, Yuma Sector Agents recovered a small bundle from a SUAS. The bundle seized contained approximately 463 grams of Methamphetamine and has an estimated street value of nearly $3,000. On Saturday night, individuals in Mexico launched a drone that entered U.S. airspace and dropped multiple packages. Responding agents were able to retrieve the ten packages dropped by the drone. The contents of the packages weighed approximately 11 kilograms of a substance that tested positive for cocaine.

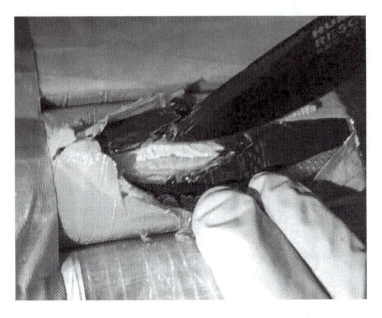

Agents seized 11 kg of cocaine that was dropped from a drone

The approximate street value of the cocaine seized is $306,000. Yuma Sector agents detected another drone on Sunday evening. Yuma Sector Border Patrol Agents recovered two of the three drones in these encounters.

Drug Trafficking Organizations (DTO's) and Transcontinental Criminal Organizations (TCO's) will go to great lengths to smuggle their narcotics. DTO's and TCO's continue to attempt various methods of smuggling narcotics. Yuma Sector Border Patrol Agents continue to remain vigilant in order to thwart and discourage these attempts.

Endnotes

[1] Release Date: 4 May 2020, https://www.cbp.gov/newsroom/local-media-release/yuma-sector-agents-intercept-narcotics-dropped-drones-0.

Addendum

Additional imagery related to the incident:

Source: U.S. Customs and Border Protection Photo. Published in "Border Patrol agents intercept 2 drones carrying narcotics in Arizona." *KTAR News.* 5 May 2020, https://ktar.com/story/3129263/border-patrol-agents-intercept-2-drones-carrying-narcotics-in-arizona/.

Chapter 15

Mexican Cartel Tactical Note #45: Drones and Explosives Seized in Puebla, Mexico by Fiscalía General de la República (FGR) and Secretaría de la Defensa Nacional (SEDENA)

David A. Kuhn, Robert J. Bunker, and John P. Sullivan

Initially Published in Small Wars Journal on 21 May 2020

Search and seizure operations were conducted by Mexican federal agencies in San Andrés Cholula, Puebla state on 25-26 April 2020 due to an anonymous tip. These operations resulted in the seizure of a 'terrorist arsenal' (*arsenal terrorista*) of weapons and equipment including the components required to make weaponized drones. This arsenal is thought to be linked to ongoing organized crime bombings in Guanajuato state—with possible Cártel de Santa Rosa de Lima (CSRL) ties—and represents one more data point concerning the growing weaponized drone capabilities of the Mexican cartels and criminal gangs.

Agencia de Investigación Criminal (AIC) agents of the
Fiscalía General de la República (FGR) raid a residence in
San Andrés Cholula containing explosives and weaponry.
FGR Photo (ESPECIAL) Reposted at: https://www.milenio.
com/policia/indaga-fgr-terrorism-hallar-c4-puebla.

Key Information: Rubén Mosso, "FGR asegura explosivos y drones en
Puebla; investiga terrorismo." *Milenio.* 3 May 2020, https://www.milenio.
com/policia/inicia-fgr-pesquisa-delito-finalidad-cometer-terrorismo:

> La Fiscalía General de la República (FGR) inició
> una carpeta de investigación por presuntos delitos de
> **delincuencia organizada** con la finalidad de cometer
> terrorismo, ya que tras diversos cateos en Puebla aseguró
> la sustancia explosiva conocida como C4, material que
> posiblemente tiene relación en ataques como el ocurrido
> en mrzo en Guanajuato un coche bomba explotó en
> Celaya…
> Durante el 25 y 26 de abril, personal ministerial de
> la FGR, con el apoyo de la Coordinación de Métodos

de Investigación de la institución y de la Secretaría de la Defensa Nacional, realizó varios cateos.

En estos se halló sustancia explosiva conocida como C4, dictaminada así por peritos de la Coordinación General de Servicios Periciales de la Fiscalía; pólvora, esferas conocidas como crisantemos, tres drones y dos controles para vehículo volador no tripulado (dron).

Asimismo, un arma de fuego corta y una larga; tres cargadores, 517 cartuchos de diversos calibres y 100 casquillos también de distintos calibres. Se aseguraron además siete teléfonos celulares y 53 tarjetas SIM; tres equipos de radiocomunicación, tarjetas de circuitos integrados, una caja con pirotecnia, así como cableado y distinta documentación, entre otros objetos.

Key Information: Rubén Mosso, "Indaga FGR terrorismo tras hallar C4 en Puebla." *Milenio.* 4 May 2020, https://www.milenio.com/policia/inicia-fgr-pesquisa-delito-finalidad-cometer-terrorismo:

> …El pasado 22 de abril, la Subprocuraduría Especializada en Investigación de Delincuencia Organizada, inició la indagatoria a raíz de una denuncia anónima en la que se mencionaron diversos hechos que pudieran constituir el delito de delincuencia organizada con la finalidad de cometer terrorismo y realizó el cateo de tres domicilios en el municipio de San Andrés Cholula, Puebla, donde halló el explosivo, armas, cartuchos y equipo de comunicación…

Key Information: "Aseguran en Puebla drones y explosivos para realizar actos terroristas." *Telefono Rojo.* 3 May 2020, https://telefonorojo.mx/aseguran-en-puebla-drones-y-explosivos-para-realizar-actos-terroristas/:

> CIUDAD DE MÉXICO.— Elementos de la Fiscalía General de la República (FGR) aseguraron drones y

explosivos en tres cateos en San Andrés Cholula, Puebla, presumiblemente para realizar actos terroristas.

De acuerdo a un comunicado, desde el pasado 22 de abril, la Fiscalía General de la República (FGR), a través de la Subprocuraduría Especializada en Investigación de Delincuencia Organizada (SEIDO), inició investigaciones con motivo de una denuncia anónima presentada el mismo día, en la que se mencionaron diversos hechos que pudieran constituir el delito de delincuencia organizada con la finalidad de cometer terrorismo.

El 25 de abril del año corriente, a petición del Ministerio Público Federal (MPF), un Juez de Control del Centro Nacional de Justicia Especializado en Control de Técnicas de Investigación, Arraigo e Intervención de Comunicaciones, autorizó la realización del cateo en tres domicilios ubicados en San Andrés Cholula, Puebla...

Key Information: Edmundo Velázquez, "FGR investiga si el Cártel de Santa Rosa de Lima maquilaba artefactos explosivos en San Pedro Cholula." *Periodico Central*. 1 May 2020, https://www.periodicocentral.mx/2020/pagina-negra/delincuencia/item/9079-fgr-investiga-si-el-cartel-de-santa-rosa-de-lima-maquilaba-artefactos-explosivos-en-san-pedro-cholula:

La Fiscalía General de la República, a través de la Subprocuraduría Especializada en Investigación de Delincuencia Organizada (SEIDO), investiga la relación del Cártel de Santa Rosa de Lima con la fabricación de supuestos explosivos en San Pedro Cholula...

Al parecer los materiales que fueron encontrados en San Pedro Cholula coinciden o son similares a los que se usaron para la explosión ocurrida el pasado 7 de abril, en Celaya, Guanajuato, un auto con explosivos detonó justo antes del puente Tres Guerras. El fuego alcanzó gran parte de un terreno baldío. De forma preliminar no se reportaron personas heridas.

Además, el 9 de marzo se dio el incendio de un vehículo en el estacionamiento de las instalaciones de la Feria de Celaya, que funciona como cuartel de la Guardia Nacional, en la comunidad de Los Mancera, en Celaya, Guanajuato.

Key Information: "FGR secures bombs and drones for terrorism, in Cholula." *EN24.* 4 May 2020, https://en24.news/en/2020/05/fgr-secures-bombs-and-drones-for-terrorism-in-cholulahtml:

> The Attorney General of the Republic (FGR) disclosed the balance of the operations carried out last weekend in different parts of the municipality of San Andrés Cholula, where he made sure **explosive material** type C4, firearms, chargers, homemade bombs and drones.
>
> Through a statement, it was reported that on April 22, through the Specialized Office of the Special Prosecutor for Organized Crime Investigation (SEIDO), the investigations began on the occasion of an anonymous complaint filed the same day for the crime of organized crime for the purpose of terrorism.
>
> Subsequently, on April 25 at the request of the Federal Public Ministry, a Control Judge of the National Center for Justice Specialized in Control of Investigative Techniques, Rooting and Communications Intervention, authorized the carrying out of searches in three addresses located in San Andrés Cholula…

Who: Organized crime members—likely a Cártel de Santa Rosa de Lima (CSRL) Cell but no official announcement from the Fiscalía General de la República (FGR) has been released.

What: Three drones, two controllers, explosives (C4, gunpowder, and mortar bomb fireworks known as 'chrysanthemums'), seven cell phones, fifty-three SIM cards, three radio communications sets, integrated circuit boards, wiring, documents, two firearms, and ammunition were seized. No detainees were reported being arrested in the raids.

When: 25-26 April (Saturday and Sunday) 2020.

Where: Three addresses in the municipality of San Andrés Cholula, Puebla state, Mexico were raided.

Why: Searches and seizures were conducted by the Fiscalía General de la República (FGR; Attorney General) along with the Secretaría de la Defensa Nacional (SEDENA; Army) as part of an ongoing investigation. These operations were linked to investigations related to earlier bombing attacks in Guanajuato state with the raid of the 'terrorist arsenal' seized resulting from an anonymous tip.

Analysis

Derived from the key information highlighted in this note, Mexican federal authorities raided three residences in San Andrés Cholula, Puebla state on 25-26 April 2020 resulting from an anonymous tip. The tip had been provided on 22 April with search warrants authorized by a Federal judge on 25 April with the raids being conducted on 25-26 April. Those individuals targeted in the raids—who were not reported to have been captured—were suspected of being linked to recent organized crime bombings in Guanajuato state that have included both improvised explosive devices (IEDs) and crude anti-personnel car bombs (e.g. VBIEDs) used primarily meant to ward off Federal agents rather than killing or injuring them.[1] One of the specific links mentioned was to the March 2020 Celaya car bombing tied to the type of explosive, which was C4, utilized.[2]

An operational cell of one of the perpetrators of those incidents—the Cártel de Santa Rosa de Lima (CSRL)—is suspected as being the target of these raids. Numerous items and artifacts were seized from the residences raided by the Fiscalía General de la República

(FGR) and the Secretaría de la Defensa Nacional (SEDENA) including firearms, ammunition, documents, cell phones and SIM cards, and radio communication sets. Additionally, the components required to construct weaponized drones were also seized during the raids. These components included three drones, two controllers, wiring, integrated circuit boards, and a cache of explosives (including C4, gunpowder, and mortar bomb fireworks).[3]

In regards to the weaponized drone components seized, the Fiscalía General de la República (FGR) released the following message along with these four images on their Twitter account on 3 May 2020:

Source: Fiscalía General de la República, FGR
México (@FGRMexico), *Twitter.* 3 May 2020,
https://twitter.com/FGRMexico/status/1257016365287911426?s=20.

The text in the tweet translated from Spanish are as follows:

The [Fiscalía General de la República] #FGR secured explosives through searches in #Puebla with the support of the [Coordinación de Métodos de Investigación] #CMI and the [Secretaría de la Defensa Nacional] @SEDENAmx; and continue the investigation. In the searches it was possible to secure radio communication equipment, drones and firearms, among other objects.

More detailed screen shots of the released images follow:

Image 1. Source: FGR México (@FGRMexico), *Twitter*. 3 May 2020, https://twitter.com/FGRMexico/ status/1257016365287911426/photo/1.

Image 2. Source: FGR México (@FGRMexico), *Twitter*. 3 May 2020, https://twitter.com/FGRMexico/ status/1257016365287911426/photo/2.

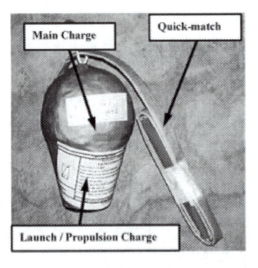

Image 2A – Professional 4-1/2-in. Mortar shell.
For Forensic Purposes. Source: David A. Kuhn.

Image 3. Source: FGR México (@FGRMexico), *Twitter*. 3 May 2020,
https://twitter.com/FGRMexico/
status/1257016365287911426/photo/3.

Image 4. Source: FGR México (@FGRMexico), *Twitter*. 3 May 2020, https://twitter.com/FGRMexico/ status/1257016365287911426/photo/4.

A basic analysis suggest the following elements of information can be obtained from this imagery:

***Image 1*:** The two drones in the foreground are DJI Inspire 2 quadcopters, while the left rear quadcopter drone is a DJI Phantom 2. This is one of the earlier models in the DJI Phantom series. The Inspire 2 quadcopters, however, are current production models and a significant step-up in price than the Phantom series. The edges of two controllers can be seen in background at the top of the photo. A circuit board has been installed to the top of each drone with wires emanating from it along with an antenna.

The reader will note that all three drones in Image No. 1 are equipped with pylons (payload) that are mounted perpendicular to the axis of normal flight. Based upon the drone dimensions, these pylons are approximately 10-inches in length. Looking at the items seized, it appears that the pyrotechnic shells that are shown in Image No. 2 are the target payload that was destined to be carried by these drones.

Image 2: The items shown within the box in Image No. 2 are the top sections of five 4-inch professional fireworks mortar shells. These shells are designed to be launched from a ground mounted mortar tube and explode at an altitude from 250-ft. to slightly greater that 400-ft. above ground level in a colored *star-burst* pattern. The diameter of the star-burst of the 4-inch shell will be approximately 350-feet (106.6-meters).

Image 2A shows the full profile of a typical 4-in. fireworks shell. The truncated shape below the sphere houses a black powder charge that propels the shell from the mortar. It is likely that the operators would ultimately remove this charge completely as it would interfere with potential targeting. Attached to each of the five shells is a 3-ft. length of *quick-match*. While the quick-match is similar in appearance to a fuze, its burn rate is much faster. Normal fireworks time fuze burns at a general rate of .25 inches per second or 12 ft./ min. Quick-match, however, burns at a very fast rate of 120-inches per second (7,200-ft./ sec.). This burn rate is designed to speed up ignition during professional fireworks displays while providing a safe firing distance to the operator igniting the guns (mortars).

Attached to each of the five shells is a 3-ft. length of *quick-match*. While the quick-match is similar in appearance to a fuze, its burn rate is much faster. Normal fireworks time fuze burns at a general rate of .25 inches per second or 12 ft./ min. Quick-match, however, burns at a very fast rate of 120-inches per second (7,200-ft./sec.). This burn rate is designed to speed up ignition during professional fireworks displays while providing a safe firing distance to the operator igniting the guns (mortars).

Image 3: The box that is shown in Image No. 3 appears to be a shipping container for pyrotechnics similar to those that are described above. Enhancement of the photograph revealed both English and Chinese characters printed on the packing/sealing tape. Some of the pixelated area of the photo are Customs declarations placards in addition to other shipping identification on the box. Enhanced photographic analysis of the type of cardboard used in the construction of the box itself indicates that it was produced in either China, Taiwan, or Korea.

The wood shavings within the box are consistent with the type of packing that is normally used for professional fireworks shipments.

Image 4: The circuit board that has been used here is a stock circuit package that is normally used as an on-board command circuit for the operation of an individual quadcopter style drone. It appears that its configuration is in an auxiliary command circuit role to both deploy and potentially begin initiation of a payload; and will draw its operational power from the drone's primary battery or from one of the two Ni-Cad. Battery packs shown in the foreground of Image No. 1.

Since this circuit is originally designed to be encased, the power leads must be secured outside of the fuselage. The communication antenna is attached directly to the BNC, or similar type connector, on the board itself. The operational frequency of this auxiliary board will also be in the low gigahertz range. One inherent weakness of this type of configuration is that it must stay dry. Any moisture, including very light rain, would result in a malfunction.

Analysis

It appears in all respects that these three drones, as configured with the auxiliary C^2 circuits and the attached pylons, were in the process of being set up to drop one or more of the 4-inch pyrotechnic shells on an unknown target. The process of configuring the drones for this type of operation requires fuzing, method of stores release (pylon), and command and control that will involve overlapping frequencies. In the case of frequency assignment, there are already two frequencies present in the operation of each drone. Command/control and video is transmitted over 2.4 GHz. There is additionally a 5.8 GHz 'Bluetooth' type link that is used to control the camera and gimble. The operator of these drones is now adding on an auxiliary C^2 board that operates on an additional frequency, the bandwidth of which is likely to be quite close to the operational frequency of the drone that it is attached to.

Keeping the frequencies from interfering with one another is just part of the overall technical burden. Additionally, there is a CG (center of gravity) problem to solve in attaching more than one payload package to each drone. While the drone's on-board microprocessor will be able to correct flight attitude for some level of CG anomalies, it cannot compensate for extreme disparity in CG within the central axis of the aircraft. There are additional technical issues that will not be addressed here due to security reasons.

The important takeaway, however, is that there are ways of overcoming all of these problems and it appears that the operators would not have equipped all three drones with auxiliary circuits and stores pylons were they not confident that they had solved the majority of them in a reliable manner.

All of the evidence present indicates that the designers/operators were planning to use these drones and their payload to communicate a threat to public officials (i.e., a politician, judge, prosecutor, or potentially even a member of law enforcement).

The preparation of three drones may have been completed for any of the following reasons:

1. Preparing for a simultaneous attack on three separate targets.
2. Preparing for an attack on a single target using more than one drone.
3. An attack using more than one drone, but keeping one in reserve for backup.

If you consider that the visually spectacular starburst that one would observe at altitude during a professional fireworks display would now be going off at ground level, it would certainly get someone's attention. While this payload is a pyrotechnic, it should not be taken lightly. Historically, professional fireworks shells such as these have maimed and killed a number of persons through a lack of respect and misuse.

A 4-inch shell detonating at ground level could be lethal at very close range and could also start fires if flammable material is present. Window glass could also be cracked and/or blown out from the concussion, depending upon the location of detonation.

All of the armed (payload carrying) cartel drones that have been observed previously in these regions were designed to land on, or near, a target and detonate. Hence, the drone has become an integral part of the guided weapon and will bultimately be destroyed when detonation occurs. The designers of these modified drones appear to have devised a payload delivery system, which, while technically difficult, would spare the weaponized drone for reuse in a potential future attack.

Of increasing concern related to the above seizures is the recent collateral proliferation of weaponized drone use in Mexico as a new phenomena since November 2017—with two earlier incidents confirmed[4]—along with a resumption of car bombings taking place over the last year or so, albeit within geographically limited regions of the country.[5] As the Mexican crime wars and criminal insurgencies continue unabated—with record breaking homicide rates evident in 2019 and now into 2020, even in the midst of the COVID-19 pandemic[6]—the use of such 'terrorist-like' weaponry may further signify that a more sanguine phase of conflict has now been entered. Such potentials do not appear lost on Mexican federal officials. Their choice of the term 'terrorist arsenal' (*arsenal terrorista*) related to this incident may signify that the use of 'narco terrorism' (*narcoterrorismo*) narratives to describe what is taking place is no longer as politically unacceptable as it once was.[7]

Sources

Rubén Mosso, "FGR asegura explosivos y drones en Puebla; investiga terrorismo." *Milenio*. 3 May 2020, https://www.milenio.com/policia/inicia-fgr-pesquisa-delito-finalidad-cometer-terrorismo.

Rubén Mosso, "Indaga FGR terrorismo tras hallar C4 en Puebla." *Milenio*. 4 May 2020, https://www.milenio.com/policia/inicia-fgr-pesquisa-delito-finalidad-cometer-terrorismo.

"Aseguran en Puebla drones y explosivos para realizar actos terroristas." *Telefono Rojo*. 3 May 2020, https://telefonorojo.mx/aseguran-en-puebla-drones-y-explosivos-para-realizar-actos-terroristas/.

Edmundo Velázquez, "FGR investiga si el Cártel de Santa Rosa de Lima maquilaba artefactos explosivos en San Pedro Cholula." *Periodico Central.* 1 May 2020, https://www.periodicocentral.mx/2020/pagina-negra/delincuencia/item/9079-fgr-investiga-si-el-cartel-de-santa-rosa-de-lima-maquilaba-artefactos-explosivos-en-san-pedro-cholula.

"FGR secures bombs and drones for terrorism, in Cholula." *EN24.* 4 May 2020, https://en24.news/en/2020/05/fgr-secures-bombs-and-drones-for-terrorism-in-cholulahtml.

Endnotes

[1] Alma Keshavarz and Robert J. Bunker, "Mexico: Another Car Bomb Explosion Near National Guard in Celaya, Guanajuato." *OE Watch.* May 2020: 89 and Alma Keshavarz and Robert J. Bunker, "Car Bomb Explodes Outside National Guard Headquarters in Celaya, Guanajuato." *OE Watch.* April 2020: 77. See also Robert J. Bunker, David A. Kuhn and John P. Sullivan, "Mexican Cartel Tactical Note #42: Car Bomb in Apaseo el Alto, Guanajuato with Remote Detonation IED ('Papa Bomba') Payload." *Small Wars Journal.* 7 January 2020, https://smallwarsjournal.com/jrnl/art/mexican-cartel-tactical-note-42-car-bomb-apaseo-el-alto-guanajuato-remote-detonation-ied.

[2] Rubén Mosso, "FGR asegura explosivos y drones en Puebla; investiga terrorismo." *Milenio.* 3 May 2020, https://www.milenio.com/policia/inicia-fgr-pesquisa-delito-finalidad-cometer-terrorismo. Another source also ties the C4 explosives recovered to a 7 April Celaya car bomb detonated by a bridge. See Edmundo Velázquez, "FGR investiga si el Cártel de Santa Rosa de Lima maquilaba artefactos explosivos en San Pedro Cholula." *Periodico Central.* 1 My 2020, https://www.periodicocentral.mx/2020/pagina-negra/delincuencia/item/9079-fgr-investiga-si-el-cartel-de-santa-rosa-de-lima-maquilaba-artefactos-explosivos-en-san-pedro-cholula.

[3] No imagery or further mention of the C4 explosives recovered in the raid have been released by the Fiscalía General de la República (FGR) linking the Cártel de Santa Rosa de Lima (CSRL) to the March and April 2020 Celaya, Guanajuato bombing incidents. It is assumed this is for operational security reasons to protect ongoing investigations and to protect FGR intelligence and forensics tradecraft.

[4] Robert J. Bunker and John P. Sullivan, "Mexican Cartel Tactical Note #35: Weaponized Drone/UAV/UAS Seized in Valtierrilla, Guanajuato with Remote Detonation IED ('Papa Bomba') Payload." *Small Wars Journal.* 23 November 2017, https://smallwarsjournal.com/jrnl/art/mexican-cartel-tactical-note-35 and John P. Sullivan, Robert J. Bunker and David A. Kuhn, "Mexican Cartel

Tactical Note #38: Armed Drone Targets the Baja California Public Safety Secretary's Residence in Tecate, Mexico." *Small Wars Journal*. 6 August 2018, https://smallwarsjournal.com/jrnl/art/mexican-cartel-tactical-note-38-armed-drone-targets-baja-california-public-safety.

[5] For car bombing context and past patterns of use, see Robert J. Bunker and John P. Sullivan, *Cartel Car Bombings in Mexico*. The Letort Papers. Carlisle, PA: US Army War College, Strategic Studies Institute. 16 August 2013, https://scholarship.claremont.edu/cgu_fac_pub/329/. For recent patterns of use, see Robert J. Bunker, David A. Kuhn and John P. Sullivan, "Mexican Cartel Tactical Note #42: Car Bomb in Apaseo el Alto, Guanajuato with Remote Detonation IED ('Papa Bomba') Payload." In Note 1.

[6] "Mexico set another record for homicides in 2019." *Business Insider*. 23 January 2020, https://www.businessinsider.com/mexico-set-another-record-for-homicides-in-2019-2020-1 and David Agren, "Mexico murder rate reaches new high as violence rages amid Covid-19 spread." *The Guardian*. 3 April 2020, https://www.theguardian.com/world/2020/apr/03/mexico-murder-rate-homicide-coronavirus-covid-19.

[7] See the following FGR press release on the investigation by the FGR special prosecutor for organized crime, Subprocuraduría Especializada en Investigación de Delincuencia Organizada (SEIDO) for an illustration of the use of 'terrorism' to describe criminal cartel and gang activity. "Comunicado FGR 130/20. La FGR aseguró explosivos mediante cateo en Puebla y continúa con la investigación." Fiscalía General de la República (FGR). Press Release. 3 May 2020, https://www.gob.mx/fgr/prensa/comunicado-fgr-130-20-la-fgr-aseguro-explosivos-mediante-cateo-en-puebla-y-continua-con-la-investigacion?idiom=es.

Significance: Cártel de Santa Rosa de Lima (CSRL), Cartel Technology, Drones, Improvised Explosive Device (IED), Tactics, Techniques, and Procedures (TTPs), Weaponized Drone, Weaponized Unmanned Aerial System (UAS)

Additional Reading

Robert J. Bunker, David A. Kuhn and John P. Sullivan, "Mexican Cartel Tactical Note #42: Car Bomb in Apaseo el Alto, Guanajuato with Remote Detonation IED ('Papa Bomba') Payload." *Small Wars Journal*. 7 January 2020, https://smallwarsjournal.com/jrnl/art/

mexican-cartel-tactical-note-42-car-bomb-apaseo-el-alto-guanajuato-remote-detonation-ied.

John P. Sullivan, Robert J. Bunker and David A. Kuhn, "Mexican Cartel Tactical Note #38: Armed Drone Targets the Baja California Public Safety Secretary's Residence in Tecate, Mexico." *Small Wars Journal.* 6 August 2018, https://smallwarsjournal.com/jrnl/art/mexican-cartel-tactical-note-38-armed-drone-targets-baja-california-public-safety.

Robert J. Bunker and John P. Sullivan, "Mexican Cartel Tactical Note #35: Weaponized Drone/UAV/UAS Seized in Valtierrilla, Guanajuato with Remote Detonation IED ('Papa Bomba') Payload." *Small Wars Journal.* 23 November 2017, https://smallwarsjournal.com/jrnl/art/mexican-cartel-tactical-note-35.

John P. Sullivan and Robert J. Bunker, "Mexican Cartel Strategic Note No. 18: Narcodrones on the Border and Beyond." *Small Wars Journal.* 28 March 2016, https://smallwarsjournal.com/jrnl/art/mexican-cartel-strategic-note-no-18-narcodrones-border-and-beyond.

Robert J. Bunker, "Mexican Cartel Tactical Note #21: Cartel Unmanned Aerial Vehicles (UAVs)." *Small Wars Journal.* 1 August 2014, https://smallwarsjournal.com/blog/mexican-cartel-tactical-note-21.

Chapter 16

El ExMarino Miembro Activo Del CJNG

Policía Comunitaria Tepalcatepec
Initially Published in Facebook on 14 July 2020

Addendum (Anthology) Translation

Former Marine Active Member of CJNG
Tepalcatepec Community Police
Image 1
"Former Marine Active Member of CJNG".

El Chino Drones an Ex-Marine military grade explosives specialist, now active in the CJNG terrorist group this character attacks masses of people or targets of the Jalisco cartel with drones loaded with C4.

Ultimately it is known that he resides in El Naranjo de Chila, birthplace of the criminal leader El Mencho.

He is also attributed with C4 attacks against El Marro in Guanajuato and in different states of the country.

(El Chino Drones [in] Blue Shirt)

Chapter 17

How Organized Crime Networks Are Using Drones to Their Advantage

Katie Jones

Initially Published in InSight Crime on 29 September 2020

Criminal groups increasingly deploying drones — for purposes ranging from surveillance to reported armed attacks of rivals — has provoked a strong government response in Mexico. But will it be enough?

To monitor and disable drones, Mexico's Defense Ministry (Secretaría de Defensa Nacional–SEDENA) plans to employ an anti-drone system costing 215.7 million pesos (about $9.6 million), according to a September 21 *El Universal*[1] report based on documents obtained by the media outlet.

The acquisition of the new technology comes about a month after a striking report that claimed[2] the Jalisco Cartel New Generation (Cártel Jalisco Nueva Generación–CJNG) had deployed drones loaded with military-grade C4 explosives in Tepalcatepec, Michoacán. The drones targeted local vigilante groups.

While the alleged equipping of drones with explosives by one of Mexico's most powerful cartels may have raised concerns among authorities, drone use by criminal groups is usually limited to surveillance.

In September, indigenous leaders in the Caru area of Brazil's northern Maranhão state said they feared their territory was being continuously monitored with drones controlled by illegal loggers and drug traffickers seeking to plant marijuana in the zone, *UOL News*[3] reported.

And drug traffickers in Guatemala employed drones to guide drug planes to clandestine landing strips hidden in Laguna del Tigre National Park, the *Washington Post*[4] reported in July.

Rangers and soldiers in the zone said[5] they could hear drones[6] fly over jungle bases about once a week. The drug planes subsequently arrived after dark, indicating that traffickers had been using the devices to survey the whereabouts of authorities to ensure a problem-free landing for their pilots.

"Their resources are infinite, and we are just trying to keep up," Juan de la Paz, a Guatemalan army colonel based in the area, told the *Washington Post*.

The Brookings Institute recently predicted that increased restriction of movement associated with the COVID-19 pandemic would only serve to speed up[7] the proliferation of drone use by crime groups.

InSight Crime Analysis

While Mexico's anti-drone plan comes on the heels of reports that such devices were weaponized by the CJNG, it appears the predominant use of drones by criminal groups is for surveillance, allowing them to monitor patrols near drug trafficking and human trafficking routes.

For example, police in Honduras claimed that the MS13 street gang used drones to escape a raid on a marijuana stash house in San Pedro Sula, *La Prensa* reported.[8] The gang also employed a video surveillance[9] network in marginal neighborhoods on San Pedro Sula's outskirts.

Human smuggling groups[10] have also deployed such technology to surveil border regions.

In April 2019, a US border patrol agent based in the El Paso sector spotted a drone[11] flying back and forth from Mexico shortly before 10 people were found attempting to cross the border illegally in the same area where the drone had been seen.

For large-scale networks and smaller organizations alike, consumer drones provide a relatively cheap, easily accessible surveillance technology.

Mark Cancian, a senior advisor at the Center for Strategic and International Studies (Centro de Estudios Estratégicos Internacionales–CSIS), told *El Confidencial*[12] earlier this year that organized crime groups have been deploying drones that are "commercially available" but "have been adapted for other purposes."

He added that "anyone can go to a shop and buy one," suggesting that drone use by such groups for surveillance purposes will only increase as potential benefits far outweigh costs.

Endnotes

[1] Noé Cruz Serrano, "Ejército va por sistema antidrones de cárteles del narco." *El Universal*. 21 September 2020, https://www.eluniversal.com.mx/nacion/ejercito-va-por-sistema-antidrones-de-carteles-del-narco.

[2] "CJNG ataca con narcodrones; autodefensas de Tepalcatepec detallan su uso." *Televisa.NEWS*. 18 August 2020, https://noticieros.televisa.com/ultimas-noticias/cjng-ataca-con-narcodrones-autodefensas-de-tepalcatepec-michoacan-detallan-su-uso/.

[3] Rubens Valente, "Drones assustam indígenas em terra alvo de madeireiros e fazendeiros no MA." *UOL*. 14 September 2020, https://noticias.uol.com.br/colunas/rubens-valente/2020/09/14/drones-invasao-terras-indigenas-maranhao.htm.

[4] Kevin Sieff, "The Guatemalan rainforest: Lush jungle, Mayan ruins and narco jets full of cocaine." *Washington Post*. 5 July 2020, https://www.washingtonpost.com/world/2020/07/05/guatemala-cocaine-trafficking-laguna-del-tigre/?arc404=true.

[5] Ibid.

[6] Kevin Sieff, "'A wildlife reserve that has become a criminal playground': The narco secrets of the Guatemalan rainforest." *The Independent*. 13 July 2020,

https://www.independent.co.uk/news/long_reads/guatemala-rainforest-narcos-cocaine-jets-drug-war-a9615746.html.

[7]   Vanda Felbab-Brown, "What coronavirus means for online fraud, forced sex, drug smuggling, and wildlife trafficking." *Brookings*. 3 April 2020, https://www. brookings.edu/blog/order-from-chaos/2020/04/03/what-coronavirus-means-for-online-fraud-forced-sex-drug-smuggling-and-wildlife-trafficking/.

[8]   "Para compartir esta nota utiliza los íconos que aparecen en el sitio." *La Prensa*. 10 February 2019, https://www.laprensa.hn/sucesos/1257909-410/autoridades-hondurenas-desmantelan-narcolaboratorio-mara-salvatrucha.

[9]   Angela Olaya, "MS13 in Honduras Gets Creative With Video Surveillance." *InSight Crime*. 13 July 2020, https://www.insightcrime.org/news/brief/ms13-video-surveillance-honduras/.

[10] Héctor Estepa, "Del narco al 'coyote': el mundo del crimen latinoamericano reinventa el uso del dron." *El Confidencial*. 1 January 2020, https://www. elconfidencial.com/mundo/2020-01-01/del-narco-al-coyote-los-nuevos-usos-del-dron-en-el-mundo-del-crimen-latinoamericano_2391035/.

[11] "El Paso Sector Border Patrol Encounters New Tactics as Smugglers Keep Sending in Families and Felons." U.S. Customs and Border Patrol. 17 April 2019, https://www.cbp.gov/newsroom/local-media-release/el-paso-sector-border-patrol-encounters-new-tactics-smugglers-keep?_ga=2.256713677.718726486.1600875406-1805667773.1599066724.

[12] Héctor Estepa, "Del narco al 'coyote': el mundo del crimen latinoamericano reinventa el uso del dron" (See Note 10).

Chapter 18

Mexican Cartel Tactical Note #46: Weaponized Drones (Aerial Improvised Explosive Devices) Deployed by CJNG in Tepalcatepec, Michoacán

Robert J. Bunker, John P. Sullivan,
David A. Kuhn, and Alma Keshavarz

Initially Published in Small Wars Journal on 5 October 2020

The *Cártel Jalisco Nueva Generación* (CJNG) is alleged to have attempted to deploy two weaponsized drones with C4 or similar type explosives against their rivals—the *Carteles Unidos* (United Cartels)—in Tepalcatepec, Michoacán on 25 July 2020. The weaponized drones were not successfully deployed and were found by a local self-defense group (*autodefensas*) known as *El grupo de autodefensa en Tepalcatepec.*

Key Information: "Jalisco cartel adopts new tactic: drones armed with C-4 explosive." *Mexico News Daily.* 18 August 2020, https://mexiconewsdaily.com/news/jalisco-cartel-adopts-new-tactic-drones-armed-with-c-4-explosive/:

A citizens' militia group in Tepalcatepec, Michoacán, reports finding two drones inside an armored car that cartel hitmen had abandoned after an attempted raid on the city, which borders Jalisco, on July 25.

The C4 was packed with ball bearings to serve as shrapnel in Tupperware-like containers that were equipped with a remote detonation system and duct-taped to the drones, militia members explained. The drones were found in a cardboard box that was soaked in blood, indicating to the militia members that whoever was intending to fly the drones was injured before they could be launched.

The new tactic represents the cartel's determination to wrest control of the western Michoacán municipality from the self-defense militia and an evolution of their air attack strategy. In April, the cartel used small planes to drop explosives on Tepalcatepec, but after authorities increased aerial surveillance in the region the CJNG opted for drones, which cannot be detected on radar.

Key Information: "CJNG ataca con narcodrones; autodefensas de Tepalcatepec detallan su uso." *Noticeros Televisa*, 18 August 2020, https://noticieros.televisa.com/ultimas-noticias/cjng-ataca-con-narcodrones-autodefensas-de-tepalcatepec-michoacan-detallan-su-uso/:

El Cártel Jalisco Nueva Generación está usando drones profesionales con C4, un explosivo de uso bélico.

"Esto es lo que dejo el Cártel Jalisco Nueva Generación, esta es la bomba C4 que iba conectada aquí, aquí en este aparato así, con esta cinta", comentó un integrante del grupo de **Autodefensa en Tepalcatepec, Michoacán.**

El pasado 25 de julio, integrantes del Cártel Jalisco Nueva Generación intentaron ingresar a Tepalcatepec donde se enfrentaron al grupo de Autodefensa de ese municipio michoacano que colinda con Jalisco.

Después de la refriega, los hombres armados de Tepalcatepec, observaron que los sicarios del cártel jalisco abandonaron dos drones y cuatro explosivos C4 en el lugar del enfrentamiento.[1]

Key Information: "CJNG usa drones con explosivos C4 y balines como forma de ataque." *El Universal.* 18 August 2020, https://www.eluniversal.com.mx/nacion/cjng-usa-drones-con-explosivos-c4-y-balines-como-forma-de-ataque:

> En entrevista, las autodefensas de la región, revelaron que tras un enfrentamiento hallaron dos **drones** y cuatro **explosivos**, por lo cual pudieron ver como la organización criminal fabrica estas nuevas armas, solo fijando las bombas con cinta metálica de uso industrial al dron.
>
> Y es que esta región de tierra caliente, colindante con Jalisco, se ha visto asolada por la organización criminal desde el 25 julio pasado, fecha en que han atacado a las **autodefensas de Tepalcatepec** para tratar de hacerse con el control del territorio.
>
> Cabe destacar que no es la primera vez que la organización criminal de **"El Mencho"** emplea esta táctica de ataque con vehículos aéreos, pues la Fiscalía General de la República (**FGR**) inició en mayo pasado una carpeta de investigación contra el CJNG.por delito de delincuencia organizada con la finalidad de cometer terrorismo.
>
> En mayo la FGR encontró, en un cateo en el estado de Puebla, varios drones y explosivo c4, mismos que presuntamente también fueron usados en Guanajuato por el Cártel.[2]

Key Information: "Drones con explosivos, la más reciente arma del CJNG para atacar desde el aire." *Infobae.* 14 August 2020, https://www.infobae.com/america/mexico/2020/08/15/drones-con-explosivos-la-mas-reciente-arma-del-cjng-para-atacar-desde-el-aire/:

El Cártel Jalisco Nueva Generación (CJNG) encontró un nuevo uso a los **drones: un arma para atacar** a quienes considera sus enemigos…

…Los drones **son cargados con el explosivo C4 y balines de metal**, conectados a un sistema de detonación a distancia.

Para los autodefensas de la localidad, los integrantes del **Cártel Jalisco Nueva Generación** aún no saben utilizar con precisión los artefactos.

Los autodefensas buscan la manera de detectar los drones que sobrevuelan el cielo de esta comunidad ubicada en la zona conocida como **Tierra Caliente**, y a 250 kilómetros de Morelia, Michoacán.[3]

Key Information: "Qué hay detrás de los supuestos drones con explosivos "asegurados" en Tepalcatepec." *Noventa Grados (90º)*. 14 August 2020, http://www.noventagrados.com.mx/seguridad/que-hay-detras-de-los-supuestos-drones-con-explosivos-asegurados-en-tepalcatepec.htm:

En redes sociales, miembros de la asociación delictiva Carteles Unidos han asegurado tener en su poder drones con explosivos que atribuyen al Cártel Jalisco Nueva Generación; este anuncio -dado hace más de un mes y que en las últimas horas ha cobrado relevancia-, parece ser una pieza más de la campaña de desinformación que realizan los grupos criminales con el afán de debilitar a sus enemigos, causando consternación entre la población…

Fue luego del ataque del CJNG a la comunidad La Estanzuela, Tepalcatepec, la noche del 25 de junio, que se conoció la primera publicación en redes sociales en la que simpatizantes de Carteles Unidos (asociación integrada por Los Viagras, Cártel de Tepalcatepec, Caballeros Templarios y Blancos de Troya) aseguraban haber localizado un dron con explosivos…

En la publicación se mostraba un dron modelo DJI Mavic 2 Zoom, con un precio de alrededor de 40 mil pesos. En él se hallaba atada con cinta metálica, una supuesta carga explosiva de C4, en lo que parece una caja rectangular color café.

En la fotografía también se apreciaba un rollo de cinta metálica.[4]

Key Information: Juan Manuel González, "Con drones, CJNG busca erradicar a rivales en Tierra Caliente." *La Silla Rota*. 12 August 2020, https://lasillarota.com/estados/con-drones-cjng-busca-erradicar-a-rivales-en-tierra-caliente-michoacan-drones-cjng-mencho/423494:

> Con **fusiles de asalto, Barrett** calibre .50 y **lanzagranadas,** el **CJNG** ataca desde tierra y, también, utiliza drones cargados con **explosivosC4** para buscar asesinar a la población. La vigilancia armada de las comunidades en esa parte colindante con **Jalisco** también acotó la irrupción del crimen organizado a sus pueblos.
>
> **La Silla Rota** dio a conocer, apenas en abril pasado, que el **CJNG** perpetraba nuevos **ataques** contra los pobladores de **Tepalcatepec.** Esos atentados eran desde avionetas, provenientes de sus centros de operaciones en **Jalisco,** desde donde lanzaban **artefactos** cargados con explosivo **C4,** cuyo acceso es exclusivo de fuerzas militares.[5]

Who: Cártel Jalisco Nueva Generación (CJNG).

What: Weaponized commercial off-the-shelf drone; **Mavic 2 Zoom,** allegedly armed with C4 (or similar type) explosives and ball bearing type projectiles in plastic containers, attached to the drones with duct tape. Two drones and four IED payloads in plastic containers were recovered.

When: Saturday, 25 July 2020.

Where: Tepalcatepec, Michoacán (Tierra Caliente), Mexico.

Why: Aerial assault; attempted attack on rival cartel.

Analysis

A weaponized drone—a **Mavic 2 Zoom** quadcopter with an IED payload—was found in a field, full of stacks of tires, by *El grupo de autodefensa en Tepalcatepec* forces (a self-defense group) on 25 July 2020 in Tepalcatepec, Michoacán.[6] This resulted in their then discovering an abandoned armored vehicle, belonging to *Cártel Jalisco Nueva Generación* (CJNG) operatives. Within that vehicle a bloody cardboard box containing another Mavic 2 Zoom quadcopter and three more 'IED payloads' consisting of clear plastic boxes containing apparent C4 explosives and industrial ball bearings (functioning as shrapnel) was secured. Apparently, the CJNG operates were involved in an attack upon competing *Carteles Unidos* (United Cartels) personnel but (it is surmised) were forced to abandon their mission due to their injuries and/or vehicular inoperability.[7]

Video imagery of the captured weaponized drones and IED components can be found at:

- Juan Manuel Gonzalez, "Con drones, CJNG busca erradicar a rivales en Tierra Caliente." *La Silla.* 12 August 2020, https://lasillarota.com/estados/con-drones-cjng-busca-erradicar-a-rivales-en-tierra-caliente-michoacan-drones-cjng-mencho/423494 [38 Second Video].
- "VIDEO | CJNG se moderniza y usa drones para realizar sus ataques." *La Voz de Michoacán.* 19 August 2020, https://www.lavozdemichoacan.com.mx/pais/cjng-se-moderniza-y-usa-drones-para-realizar-sus-ataques/ [39 Second Video].

The video imagery shows:

a. One gray Mavic 2 Zoom quadcopter (Drone 1) with an 'IED payload' (Payload 1) found in a clear plastic container with a clear top/blue locking latches duct taped to it.

b. Another gray Mavic 2 Zoom quadcopter (Drone 2) with no 'IED payload' attached.

c. A second 'IED payload' (Payload 2) found in a clear top/blue locking handles clear plastic container.

d. Two additional 'IED payloads' (Payload 3 & 4) found in clear plastic containers with green lids and sealed with masking and/or electrical tape.

Specific information pertaining to the IED design and effects related to drone weaponization—along with imagery forensics—follows[8]:

There are four (4) plastic containers that were discovered at the site. One of these containers is shown in Image 1 with the payload duct taped to the drone's fuselage. Two of the containers discovered (with the blue locking latches) are general storage containers.

Image 1: Payload attached to drone's fuselage. (Source: *La Silla Rota*, used with permission. See Note 8.)

Image 2: Payload containers (Source: *La Silla Rota*, used with permission. See Note 8.)

These particular containers have an internal volume of 0.3-liters (10.14 oz.). The two remaining containers with the green lids are unknown generic, sealable food containers that possess a slightly lower internal volume. The payloads of all of the containers, where visible, contain industrial ball bearings of at least two sizes and a plastic explosive that has been identified in a number of reports and videos as Composition C-4. However, the color of the explosive within the containers does not match the color any of the normal C-4 composition grades in production (Image 3).

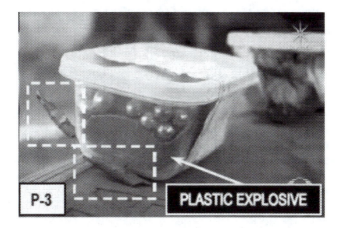

Image 3: Interior contents of payload containers (Source: *La Silla Rota*, used with permission. See Note 8.)

The color does, however, match very closely to one of the grades of Semtex explosives. Composition C-4 is somewhat prevalent throughout Mexico and the auto-defense force providing the interview and information for the videos may have assumed that it was C4 rather than Semtex. The distinction, if accurate, however, may have far more value from an intelligence viewpoint as both Composition C-4 and several grades of Semtex have very similar performance characteristics.

The shrapnel that the bomb makers are using here are industrial ball bearings that have been salvaged from full bearing assemblies. Industrial ball bearings, by their nature, are constructed of precision, hardened steel. The bearings within the containers are of various sizes including 3/8-inch (9.52mm) and ½-inch (12.7mm). This package, coupled with the plastic explosive, would be devastating at a reasonably close range.

It appears that the plastic explosive may have been shaped (contoured) within the containers prior to inserting the ball bearings (Image 3).

The Mavic 2 Zoom Drone shown in Image 1 has the payload package secured to it via duct tape. This creates a relatively simple, remote operated, attack system. The effective deployment range of this drone for such an attack, however, is limited. While the Mavic 2 generally has a maximum flight range of approximately 12-miles (with no additional payload factored in), its maximum video transmission range about is 5-statute miles (8.04 km.) under relatively good weather conditions. The operators will require that video to locate their intended target. The payload package, if delivered as configured as in the first photograph, will also destroy the drone.

One of the containers shown in Image 4 been cut away on one end with the lid still intact. This container also has black electrical tape wrapped over part of the lid and body. The ball bearings and plastic explosive appear to be packaged within a small gray retail type plastic bag. It has been placed within the container opening and is held there apparently by friction only. It is possible that the cartel was conducting experiments at this site in order to determine the feasibility of air dropping the payload by pitching the drone forward over a potential target; thus saving the drone for future attack.

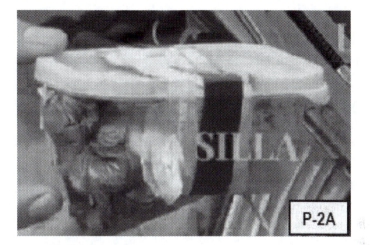

Image 4: Possible experimental IED load (Source: *La Silla Rota*, used with permission. See Note 8.)

The reader will also note that that the format of the container that is attached to the drone in Image 1 is again different that the other containers. It appears that the cartel operatives have stabilized (fixed) the ball bearing payload within the container using what appears to injectable foam that will harden enough to retain the contents in an exact position.

Conclusion

A few years ago, it would have been impossible for even a similar sized mil-spec drone to lift payloads of the approximate weight that we are seeing here, however, now they have become readily available to virtually anyone. The skill curve that exists in order to successfully maneuver modern drones and, in particular the Mavic 2 series, is extremely low. On-board microprocessors and embedded flight control sensors do the majority of the work. Drones possessing this type of technology are already in a special asymmetric threat category that requires special defense measures and early interdiction to be successful.

Criminal cartels in Mexico—in particular the CJNG—are continuing to experiment with the weaponization of commercial, off-the-shelf (COTS) drones to target their adversaries. In this incident, the attack was interrupted or aborted. It can be expected that cartels will continue their experiments with weaponized aerial drones and that this experimentation will yield more sophisticated devices and refined TTPs (tactics, techniques, and procedures). In recognition of the increasing cartel weaponized drone threat in Mexico, SEDENA (*Secretaría de Defensa Nacional*), per information obtained by *El Universal* in September 2020, now "plans to employ an anti-drone system costing 215.7 million pesos (about $9.6 million)" to monitor and disable such systems.[9]

Sources

"CJNG ataca con narcodrones; autodefensas de Tepalcatepec detallan su uso." *Noticeros Televisa*. 18 August 2020, https://noticieros. televisa.com/ultimas-noticias/cjng-ataca-con-narcodrones-autodefensas-de-tepalcatepec-michoacan-detallan-su-uso/.

"CJNG usa drones con explosivos C4 y balines como forma de ataque." *El Universal*. 18 August 2020, https://www.eluniversal.com.mx/nacion/cjng-usa-drones-con-explosivos-c4-y-balines-como-forma-de-ataque.

"Drones con explosivos, la más reciente arma del CJNG para atacar desde el aire." *Infobae*. 14 August 2020, https://www.infobae.com/america/mexico/2020/08/15/drones-con-explosivos-la-mas-reciente-arma-del-cjng-para-atacar-desde-el-aire/.

Juan Manuel González, "Con drones, CJNG busca erradicar a rivales en Tierra Caliente." *La Silla Rota*. 12 August 2020, https://lasillarota.com/estados/con-drones-cjng-busca-erradicar-a-rivales-en-tierra-caliente-michoacan-drones-cjng-mencho/423494.

"Jalisco cartel adopts new tactic: drones armed with C-4 explosive." *Mexico News Daily*. 18 August 2020, https://mexiconewsdaily.com/news/jalisco-cartel-adopts-new-tactic-drones-armed-with-c-4-explosive/.

"Qué hay detrás de los supuestos drones con explosivos "asegurados" en Tepalcatepec." *Noventa Grados (90°)*. 14 August 2020, http://www.

noventagrados.com.mx/seguridad/que-hay-detras-de-los-supuestos-drones-con-explosivos-asegurados-en-tepalcatepec.htm.

Significance: Aerial Improvised Explosive Devices (A-IED), *Autodefensas*, C4, *Cártel Jalisco Nueva Generación* (CJNG), *Carteles Unidos*, Drones, Narcodrones, Unmanned Aerial Systems (UAS), Unmanned Aerial Vehicles (UAVs), Michoacán.

Endnotes

[1] In English, the title reads: "CJNG attacks with narcodrones; Tepalcatepec self-defense groups detail their use." The text reads: "The *Cártel de Jalisco Nueva Generación* (Jalisco New Generation Cartel) [CJNG] is using professional drones with C4, an explosive for military use." ... "'This is what the Jalisco New Generation Cartel left, this is the C4 bomb that was connected here, here in this device like this, with this tape,' commented a member of the Self-Defense group in Tepalcatepec, Michoacán." ... "On July 25, members of the Jalisco New Generation Cartel tried to enter Tepalcatepec where they confronted the Self-Defense group of that Michoacán municipality bordering Jalisco." ... "After the skirmish, the armed men of Tepalcatepec observed that the Jalisco cartel hitmen abandoned two drones and four C4 explosives at the scene of the confrontation."

[2] In English, the title reads: "CJNG uses drones with C4 explosives and pellets as a form of attack." The text reads: "In an interview, the region's self-defense groups revealed that after a confrontation they found two drones and four explosives, so they could see how the criminal organization manufactures these new weapons, only fixing the bombs with metal tape for industrial use to the drone." ... "And it is this hot land region, adjacent to Jalisco, that has been ravaged by the criminal organization since last July 25, the date on which they attacked the self-defense groups in Tepalcatepec to try to gain control of the territory." ... "It should be noted that this is not the first time that "El Mencho's criminal organization has used this aerial vehicle attack tactic, since the Fiscalía General de la República (Attorney General's Office) (FGR) opened an investigation against the CJNG last May for the offense of organized crime for the purpose of committing terrorism." ... "In May, the FGR found, in a search in the state of Puebla, several drones and a C4 explosive, which were allegedly also used in by the cartel in Guanajuato."

[3] In English, the title reads: The text reads: "The *Cártel de Jalisco Nueva Generación* (Jalisco New Generation Cartel) (CJNG) found a new use for drones: a weapon to attack those they consider their enemies" ... "The drones are loaded with the C4 explosive and metal pellets, connected to a remote detonation system." ...

"For the town's *autodefensas* [self-defense groups] the members of the Jalisco New Generation Cartel still do not know how to use the devices with precision." … "The self-defense groups are looking for a way to detect the drones that fly over the sky of this community located in the area known as Tierra Caliente, and 250 kilometers from Morelia, Michoacán."

[4] In English the title reads: "What is behind the alleged drones with explosives 'secured' in Tepalcatepec." The text reads: "In social networks, members of the criminal association *Carteles Unidos* have claimed to have in their possession drones with explosives that they attribute to the Jalisco *Cártel de Jalisco Nueva Generación* (Jalisco New Generation Cartel) [CJNG] This announcement – given more than a month ago and which has gained relevance in the last few hours – seems to be one more piece of the disinformation campaign carried out by criminal groups with the aim of weakening their enemies, causing consternation among the population." … "It was after the CJNG attack on the La Estanzuela community, Tepalcatepec, on the night of June 25, that the first publication on social networks in which sympathizers of *Carteles Unidos* (an association made up of *Los Viagras*, *Cártel* [Tepalcatepec Cartel], Caballeros Templarios [Knights Templar] and the *Blancos de Troy* [Whites of Troy] claimed to have located a drone with explosives." … "The publication showed a **DJI Mavic 2 Zoom model drone** [emphasis added], with a price of around 40 thousand pesos. In it was tied with metal tape, an alleged explosive charge of C4, in what appears to be a rectangular brown box." … "The photograph also showed a roll of metal tape [duct tape]".

[5] In English, the title reads: "With drones, CJNG seeks to eradicate rivals in Tierra Caliente." The text reads: "With assault rifles, Barrett .50 caliber [rifles] and grenade launchers, the CJNG attacks from the ground and also uses drones loaded with C4 explosives seeking to assassinate the population. The armed community surveillance [*autodefensas*] in parts adjacent to Jalisco also limited the irruption of organized crime in their towns." … "La Silla Rota announced, just last April, that the CJNG was carrying out new attacks against the inhabitants of Tepalcatepec. These attacks were from airplanes, coming from their operations centers in Jalisco, from where they were launching devices loaded with C4 explosives, exclusively available to military forces."

[6] Juan Manuel González, "Con drones, CJNG busca erradicar a rivales en Tierra Caliente." *La Silla Rota*. 12 August 2020, https://lasillarota.com/estados/con-drones-cjng-busca-erradicar-a-rivales-en-tierra-caliente-michoacan-drones-cjng-mencho/423494. This reporting is in variance with one of the incident images which shows both drones in the field together.

[7] Ibid. No imagery of the bloody cardboard box in which one of the drones and three of the IED payloads were discovered—or for that matter the abandoned armored SUV—exists.

[8] Special thanks to Jorge Ramos, Director Editorial, Grupo La Silla Rota for

providing permission in an email on 22 August 2020 for us to utilize their video imagery of the incident for this tactical note and follow on analytical and book products.

[9] Katie Jones, "How Organized Crime Networks Are Using Drones to Their Advantage." *InSight Crime.* 29 September 2020, https://www.insightcrime. org/news/brief/drones-narcotrafficking-surveillance/. See also, Noé Cruz Serrano, "Ejército va por sistema antidrones de cárteles del narco." *El Universal.* 21 September 2020, https://www.eluniversal.com.mx/nacion/ ejercito-va-por-sistema-antidrones-de-carteles-del-narco.

For Additional Reading

David A. Kuhn, Robert J. Bunker, and John P. Sullivan, "Mexican Cartel Tactical Note #45: Drones and Explosives Seized in Puebla, Mexico by Fiscalía General de la República (FGR) and Secretaría de la Defensa Nacional (SEDENA)." *Small Wars Journal.* 21 May 2020, https://smallwarsjournal.com/jrnl/art/mexican-cartel-tactical-note-45-drones-and-explosives-seized-puebla-mexico-fiscalia.

John P. Sullivan, Robert J. Bunker, and David A. Kuhn, "Mexican Cartel Tactical Note #38: Armed Drone Targets the Baja California Public Safety Secretary's Residence in Tecate, Mexico." *Small Wars Journal.* 6 August 2018, https://smallwarsjournal.com/jrnl/art/mexican-cartel-tactical-note-38-armed-drone-targets-baja-california-public-safety.

Robert J. Bunker and John P. Sullivan, "Mexican Cartel Tactical Note #35: Weaponized Drone/UAV/UAS Seized in Valtierrilla, Guanajuato with Remote Detonation IED ('Papa Bomba') Payload." *Small Wars Journal.* 23 November 2017, https://smallwarsjournal.com/ jrnl/art/mexican-cartel-tactical-note-35.

John P. Sullivan and Robert J. Bunker, "Mexican Cartel Strategic Note No. 18: Narcodrones on the Border and Beyond." *Small Wars Journal.* 28 March 2016, https://smallwarsjournal.com/jrnl/art/ mexican-cartel-strategic-note-no-18-narcodrones-border-and-beyond.

Robert J. Bunker, "Mexican Cartel Tactical Note #21: Cartel Unmanned Aerial Vehicles (UAVs)." *Small Wars Journal.* 1 August 2014, https://smallwarsjournal.com/blog/mexican-cartel-tactical-note-21.

Chapter 19

Yuma Agents Detect Cross-Border Drone Smuggling Narcotics

U.S. Customs and Border Protection

Initially Published in CBP Newsroom on 9 November 2020 (Release Date)[1]

YUMA, Ariz. – Yuma Sector agents assisted in the arrest of two U.S. citizens on Thursday after they picked up multiple packages of narcotics that were dropped by a drone coming from Mexico.

Yuma Station agents detected a small-unmanned aircraft system making multiple trips into the United States from Mexico near San Luis at 1:45 a.m. Agents observed the drone dropping multiple packages, which were later determined to contain methamphetamine, and detained the two subjects near the residence where the packages were dropped.

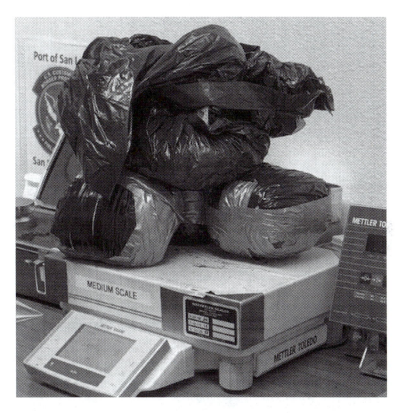

Yuma Sector Border Patrol agents assisted in the arrest of two U.S. citizens for picking up narcotics dropped by a drone from Mexico.

The Yuma County Narcotics Task Force served a search warrant on the residence and a subsequent search resulted in the seizure of 24.9 pounds of methamphetamine and a .357 magnum revolver. The estimated street value of the methamphetamine was over $225,000.

The task force took custody of the male and female subjects, as well as the narcotics and handgun.

Endnotes

[1] Release Date: 9 November 2020, https://www.cbp.gov/newsroom/local-media-release/yuma-agents-detect-cross-border-drone-smuggling-narcotics?_ga=2.215043859.584795156.1631411989-349536333.1631411989.

Chapter 20

Mexican Cartel Tactical Note #48: Video of CJNG Engagement of Autodefensa Mounted Infantry in IAFV in La Bocanda, Michoacán

Robert J. Bunker and John P. Sullivan

Initially Published in Small Wars Journal on 30 December 2020

In the ongoing conflict in Michoacán between the *Cártel Jalisco Nueva Generación* (CJNG) and local defense forces coordinating amongst themselves and with local cartels, a recent engagement took place in which an *autodefensa* (self-defense group) improvised armored fighting vehicle (IAFV) was engaged by small arms fires. The engagement, caught on video from the perspective of the mounted infantry and reporters in the armored truck bed of the IAFV, took place in La Bocanda—a CJNG stronghold—in mid-December 2020.

Key Information: "Captan en video emboscada del CJNG contra autodefensas y reporteros." *El Universal.* 17 December 2020, https://www.eluniversal.com.mx/estados/captan-en-video-emboscada-del-cjng-contra-autodefensas-y-reporteros-0:

Los reporteros **Marco Antonio Coronel** y **Fernando Guillén**, integrantes del equipo del noticiero **En Punto de Dennise Maerker,** fueron atacados a **balazos,** luego de asistir el pasado miércoles 16 a la zona conocida como **La Bocanda,** en el municipio de **Tepalcatepec, Michoacán,** acompañados de **Autrodefensas** de la zona…

Los reporteros fueron a **La Bocanda** para documentar cómo las personas del lugar huyen, debido a la violencia que hay por los constantes enfrentamientos entre integrantes del **CJNG** y los autodenominados **Autodefensas** de Tepalcatepec.

En el reportaje narran como el **CJNG** convirtió a **La Bocanda** en su base de operaciones, y como utilizan una edificación, a la que llaman La casa baleada, como fortín para atacar desde ahí a los miembros de las Autodefensas, que se apostaron metros adelante para impedir que los del cártel sigan avanzando.

En el video del ataque, uno de los reporteros incluso enfoca la marca de un balazo que quedó detras de él, mostrando la cercanía a la que pasaron las balas de ellos. También, en el video, se ve a uno de los autodefensas herido del brazo.[1]

The video and the fortified building are documented in the following two entries.

Key Information: Marco A Coronel, "El #CártelJaliscoNuevaGeneración convirtió La Bocanda, Michoacán, en su base de operaciones." *Twitter.* 17 December 2020, https://twitter.com/marcocoronel/ status/1339787765513396226?s=20:

El #CártelJaliscoNuevaGeneración convirtió La Bocanda, Michoacán, en su base de operaciones. La comunidad es una zona de guerra. Viajamos ahí, con ayuda de las #Autodefensas, para captar el impacto a la población y fuimos emboscados. Nuestro reporte #EnPunto @NTelevisa_com.[2]

Marco A Coronel
@marcocoronel

El #CártelJaliscoNuevaGeneración convirtió La Bocanda,
Michoacán, en su base de operaciones. La comunidad es
una zona de guerra. Viajamos ahí, con ayuda de las
#Autodefensas, para captar el impacto a la población y
fuimos emboscados. Nuestro reporte #EnPunto
@NTelevisa_com

Translate Tweet

89.5K views   0:25 / 0:50

8:21 PM · Dec 17, 2020 · Twitter for Android

993 Retweets   **216** Quote Tweets   **2.2K** Likes

Autodefensa IAFV Truck Bed
Source: Tweet Courtesy of Marco A. Coronel.

Key Information: "Autodefensas se enfrentan al CJNG en Michoacán."
FOROtv at Facebook. 18 December 2020, https://www.facebook.com/
FOROtv/videos/1082440872201669/:

CJNG Fortified Building (La Casa Baleada) in La
Bocanda. Autodefensa Drone Imagery.
Source: Screen shot ForoTV at Facebook.

Key Information: "#CJNG EMBOSCA A AUTODEFENSAS Y REPORTEROS EN #LABOCANDA VIDEO." *Valor Por Tamaulipas.* 18 December 2020, https://www.valorportamaulipas.info/2020/12/cjng-embosca-autodefensas-y-reporteros.html:

> Circula en redes sociales el fragmento de un video, en el que miembros del Cártel Jalisco Nueva Generación, emboscan a reporteros y a un grupo de autodefensas…
>
> …La Bocanda, Michoacán, se ha convertido en la base de operaciones del CJNG y la comunidad se ha convertido en un campo de guerra.[3]

Key Information: "CJNG ataca a reporteros que acompañaban a autodefensas de Michoacán." *Vanguardia.* 19 December 2020, https://

Los reporteros se encontraban documentando la realidad que se vive en el municipio cuando fueron atacados por CJNG

Un par de reporteros del noticiero *En Punto* de **Televisa** y miembros de autodefensas locales fueron atacados por miembros del **Cártel Jalisco Nueva Generación** (CJNG) el pasado miércoles 16 de diciembre en **La Bocanada, Michoacán**…

…No obstante, al momento en que el grupo llegó al lugara que se encuentra en poder del **CJNG**, los delincuentes los recibieron a balazos, por lo que los autodefensas repelieron el ataque de los agresores; no obstante, uno de ellos resultó con una herida en el brazo.[4]

Who: Gunmen armed with infantry small arms belonging to the *Cártel Jalisco Nueva Generación* (CJNG – Jalisco New Generation Cartel).

What: The CJNG gunmen fire upon a IAFV containing mounted infantry—belonging to the *Autodefensas de Tepalcatepec* (Self-Defense Forces of Tepalcatepec)—in the bed of the armored heavy truck along with two reporters, Marco Antonio Coronel and Fernando Guillén from *En Punto de Televisa*, accompanying them.

When: Wednesday, 16 December 2020 during daylight hours.

Where: La Bocanda municipality between Aguililla and Tepalcatepec, Michoacán.

Why: The CJNG gunmen engaged the *autodefensa* members (and the reporters) riding in the IAFV as a component of their ongoing operations to exert control over the region of La Bocanda. The CJNG has established a fortified position in a building called *La Casa Baleada*, which they are utilizing as an operational base. The reporters were riding with the *autodefensa* members into La Bocanda to chronicle the violence and depopulation of the area caused by ongoing CJNG activities.

Analysis

The incident videos involving the *autodefensa* IAFV and the CJNG forces firing upon it pertaining to the fortified CJNG forward base (aka *La Casa Baleada; the* 'Shot House')[5] in La Bocanda has been widely viewed on social media and in news reports. The basic 5Ws providing a short overview of the tactical action have been provided above in this note. Some context behind the incident was highlighted via recent correspondence with Marco A. Coronel who was riding inside the IAFV when it was attacked:

> La primera parte del recorrido viajamos en una de varias camionetas en carabana (se observa en el reportaje) después llegamos a una especie de campamento donde nos subieron al blindado para poder aproximarnos más a la comunidad de La Bocanda y así poder grabar con nuestro dron a distancia.[6]

Hence, the *autodefensa* group utilized the safety of their interior lines to transport the *En Punto de Televisa* reporters (Marco A. Coronel and Fernando Guillén) to their base camp near La Bocanda in a caravan of soft-skinned vehicles (i.e., vans). The videos posted on Twitter (17 December 2020) and Facebook (18 December 2020) provide imagery and an audio account of the tactical engagement itself after the reporters were loaded up into an *autodefensa* IAFV (with an accompanying security detail armed with assault rifles, a .50 Cal. Barrett rifle, and wearing ballistic armor) and set off on the reconnaissance mission. It is still unclear how many accompanying *autodefensa* IAFVs were or were not part of this mission—drone video imagery segments from the Facebook posting (see 1:35 and 2:38) first shows La Bocanda and then a column of roughly eleven *autodefensa* vehicles in the vicinity possibly involved in supporting the operation.

The incident videos portray the confusion, moments of sheer terror, and inherent danger found with such tactical engagements. In addition to sparks and sounds of impacts, some CJNG rounds are clearly seen to penetrate the IAFV's armor a few feet up to the right from Coronel's head in the Twitter (0:48-0:50) posting and later, in the Facebook (4:21) posting, an *autodefensa* can clearly be seen to have been wounded in the upper left arm.

Two image sets have been selected from the Twitter smart phone video posting uploaded by Marco A. Coronel from his vantage point via the passenger side rear section of the armored truck bed. They are of tactical interest for the following reasons:

Image Set 1. The initial image set provides perspectives on the interior components of the IAFV bed and the ISR drone. The mounted infantry section of the IAFV has cutout firing ports that can be popped out and then resealed with a cable/rebar pull back system. These firing slots were utilized during the ensuing engagement that took place. Also evident is a fixed weapons mount in the center of the bed and an armored firing 'V' slot right behind the back of the vehicle cab directed forward. While Barrett .50 Cal rifles have been stabilized by such mounts—as viewed in past cartel weaponry imagery[7]—this mount is likely meant for heavier infantry weapons such as M2 Browning ("Ma Deuce") or smaller caliber machine guns as seen in cartel arsenals.[8] The possession of such fixed mounted infantry weapons would currently put *autodefensa* IAFVs at offensive parity with deployed cartel IAFVs. The use of a drone (in this instance, a DJI Mavic Mini) for ISR purposes by the *autodefensa* unit and the fact it was launched and recovered from the bed of the IAFV (even though the vehicle appeared not to be moving in both instances; Facebook 3:09 and 3:19) is also of significance. Verification of such *autodefensa*—or, for that matter, cartel—ISR drone use is very rare and to have it utilized as a means to conduct a reconnaissance mission from an IAFV is, to our knowledge, a first.

Forward Armored Firing V Slot

Firing Port; Pop Out with Cable/Rebar Retrieval Pull Back

Fixed Weapons Mount

0:03 of 0:50 Seconds of Video

Drone; DJI Mavic Mini

0:02 of 0:50 Seconds of Video   Drone RF Controller   0:13 of 0:50 Seconds of Video

Marco A Coronel, "El #CártelJaliscoNuevaGeneración convirtió La Bocanda, Michoacán, en su base de operaciones," Twitter. 17 December 2020. *Courtesy of Marco A. Coronel.*

Image Set 1: Autodefensa IAFV Truck Bed: Lay Out and ISR Drone/Controller

Sinaloa Cartel IAFV (Bed/Cargo Box Armored Only) with .50 Cal. M2 Machine Gun on Fixed Weapons Mount in Culicán, Sinaloa in October 2019. Aborted Operation to Arrest Ovidio Guzmán López. Source: Cell Phone Social Media Video Posting (Unattributed)

Image Set 2. The second image set focuses on the Barrett .50 Cal. rifle gunman. He is wearing radio/commo gear on the left side of his head and a ballistic vest as well as carrying supplies on the front of his utility harness and in a green satchel/backpack attached to it. A shooting glove (with the fingers exposed) on his left hand is also evident while he stabilizes and aims the weapon. Whereas Barrett .50 Cal. rifles are now relatively common in the arsenals of the cartels,[9] they are more seldom seen deployed by the *autodefensas* as are RPGs (rocket propelled grenades).

Marco A Coronel, "El #CártelJaliscoNuevaGeneración convirtió La Bocanda, Michoacán, en su base de operaciones." Twitter. 17 December 2020. *Courtesy of Marco A. Coronel.*

Image Set 2: Autodefensa IAFV Truck Bed: Barrett .50 Cal. Rifle Gunman

The tactical action at La Bocanda in many ways provides a vignette into both the present intensity of the criminal insurgencies waging across regions of Mexico and the use of various technologies and forms of weaponry by the belligerents engaging one another. An invading cartel utilizing a fortified building as a forward strong point is reconned by a ISR drone belonging to a civilian militia (possibly compromised or not by competing cartels) launched from the bed of an armored truck. The intent of the recon mission is to provide imagery for the two reporters aboard the vehicle to help them broadcast information about the depopulation and other deprivations taking place within the local communities. The *sicarios* (assassins) of the invading cartel—the

CJNG—in response fire upon the IAFV to drive it away from their operational base.

Nowhere to be seen are Mexican federal, state, or local authorities— in fact, the immediate area appears devoid of a stabilizing law enforcement or military presence. As in other regions of Mexico, the central government's 'low political capacity' is being strained to the breaking point by the second order effects of the ongoing pandemic— including increasing budget deficits and redeployment of ground forces—now layered upon the already endemic cartel and gang conflicts. To be fair, for many years now, Michoacán has been a fragile if not failed region within Mexico. Who can forget the earlier criminal (and spiritual) insurgent activities of *La Familia Michoacana* and the *Los Caballeros Templarios* and the open societal conflict that resulted as local communities rose up against the tyranny of their neo-feudal pseudo-Christian cult practices.[10]

Still, local conditions appear to be once again getting worse with the recent CJNG invasion. Anti-IAFV trenches now sporadically cut across major highways within the state[11] and the *autodefensas* now field their own IAFVs and large caliber infantry weapons in response to the invading CJNG commandos—in essence, motorized IAFV mounted infantry units—presently attempting to overrun the local defenders. In summation, the constant conflict in sections of Michoacán, and within other states such Guanajuato Colima, and Guerrero, provide us insights into patterns of technology and weaponry use not only among the gangs and cartels but also within the *autodefensas* and similar people's militias that have emerged to contest their attempted criminal domination over local communities. Greater numbers of improvised armored fighting vehicles, heavy infantry weapons, and increasingly drones—for both ISR and combat purposes—are becoming part of the landscape of the criminal insurgencies taking place within Mexico.

Sources

"Autodefensas se enfrentan al CJNG en Michoacán." FOROtv at Facebook. 18 December 2020, https://www.facebook.com/FOROtv/videos/1082440872201669/.

"Captan en video emboscada del CJNG contra autodefensas y reporteros." *El Universal.* 17 December 2020, https://www.eluniversal.com.mx/estados/captan-en-video-emboscada-del-cjng-contra-autodefensas-y-reporteros-0.

"CJNG ataca a reporteros que acompañaban a autodefensas de Michoacán." *Vanguardia.* 19 December 2020, https://vanguardia.com.mx/articulo/cjng-ataca-reporteros-que-acompanaban-autodefensas-de-michoacan.

"#CJNG EMBOSCA A AUTODEFENSAS Y REPORTEROS EN #LABOCANDA VIDEO." *Valor Por Tamaulipas.* 18 December 2020, https://www.valorportamaulipas.info/2020/12/cjng-embosca-autodefensas-y-reporteros.html.

Marco A Coronel, "El #CártelJaliscoNuevaGeneración convirtió La Bocanda, Michoacán, en su base de operaciones." *Twitter.* 17 December 2020, https://twitter.com/marcocoronel/status/1339787765513396226?s=20.

Significance: *Autodefensas*, Barrett .50 Cal. Rifle, *Cártel Jalisco Nueva Generación* (CJNG), Drone(s), Improvised Armored Fighting Vehicles (IAFV), Intelligence Surveillance, and Reconnaissance (ISR), *Narcotanques* (Narco-Tanks), Self-Defense Forces of Tepalcatepec, Unmanned Aerial System (UAS)

Endnotes

Special thanks to Marco A. Coronel, Ioan Grillo, and Conrad Dreby for their help in the preparation of this note.

[1]   In English, the title reads: "CJNG ambush against self-defense groups and reporters captured on video." The text reads: "Reporters Marco Antonio Coronel

and Fernando Guillén, members of Dennise Maerker's *En Punto* newscast team, were attacked with gunfire, after visiting the area known as La Bocanda accompanied by Self-defense groups on Wednesday 16 [December 2020], in the municipality of Tepalcatepec, Michoacán.".…"The reporters went to La Bocanda to document how the people of the place flee, due to the violence due to the constant confrontations between members of the CJNG and the self-described Self-Defense Forces of Tepalcatepec.".…"In the report they narrate how the CJNG turned La Bocanda into their base of operations, and how they use a building, which they call *La casa baleada* [the shot house], as a fort to attack the members of the Self-Defense Forces, who stationed themselves meters ahead to stop the cartel from advancing.".…"In the video of the attack, one of the reporters even focuses on the mark of a bullet that was left behind him, showing the closeness to which the bullets passed from them. Also, in the video, one of the self-defense groups is seen with injures in the arm."

[2]   In English, the Tweet reads: "The #CártelJaliscoNuevaGeneración made La Bocanda, Michoacán, its base of operations. The community is a war zone. We traveled there, with the help of #Autodefensas, to capture the impact on the population and we were ambushed. Our report #EnPunto @NTelevisa_com."

[3]   In English, the title reads: "#CJNG Ambush *Autodefensas* [Self-Defense Forces] and Reporters in #LaBocanda Video." The text reads: "A video fragment circulates on social networks, in which members of the Jalisco New Generation Cartel ambush reporters and [members of] a self-defense group..".…"La Bocanda, Michoacán, has become the CJNG's base of operations and the community has become a battlefield."

[4]   In English, the title reads: "CJNG attacks reporters accompanying Michoacán *autodefensas* [self-defense groups]." The text reads: "The reporters were documenting the reality of the municipality when they were attacked by CJNG.".…"A pair of reporters from the *En Punto de Televisa* newscast and members of local self-defense groups were attacked by members of the Jalisco New Generation Cartel (CJNG) last Wednesday, 16 December [2020], in La Bocanada, Michoacán.".…"However, when the group arrived at the place controlled by the CJNG, the criminals received them with bullets, so the self-defense groups repelled the attack by the aggressors; however, one of them was injured in the arm."

[5]   "'La casa baleada', fortaleza del CJNG desde la que mantienen una cruda batalla contra las autodefensas de Michoacán." *Infobae*. 18 December 2020, https://www.infobae.com/america/mexico/2020/12/18/la-casa-baleada-fortaleza-del-cjng-desde-la-que-mantienen-una-cruda-batalla-contra-las-autodefensas-de-michoacan/.

[6]   In English, the passage reads "[During] the first part of the tour we traveled in one of several vans in a caravan (as seen in the report) then we arrived at a kind

of camp where they put us on the armored vehicle to be able to get closer to the community of La Bocanda and thus be able to record with our drone at a distance." Email correspondence with Marco A. Coronel, 22 December 2020.

[7] Robert J. Bunker and Byron Ramirez, Eds., *Narco Armor: Improvised Armored Fighting Vehicles in Mexico.* Leavenworth, KS: Foreign Military Studies Office, US Army Command and General Staff College. October 2013: pp. 74-75, https://community.apan.org/wg/tradoc-g2/fmso/m/fmso-monographs/197127.

[8] Ibid; Also see "'Armored division' supported Sinaloa Cartel in Culiacán." *Mexico News Daily.* 21 October 2019, https://mexiconewsdaily.com/news/armored-division-supported-sinaloa-cartel/ and "2 convoy videos were the work of Jalisco cartel's 'Elite Group:' army chief." *Mexico News Daily.* 21 July 2020, https://mexiconewsdaily.com/news/2-convoy-videos-were-the-work-of-jalisco-cartels-elite-group-army-chief/.

[9] Robert J. Bunker, "Sniping in the Mexican Criminal Insurgency." *The Counter Terrorist.* June/July 2014: pp. 30-32, 34-38, 40-42, http://www.thecounterterroristmag.com/pdf/issues/TheCounterTerrorist_JuneJuly2014.pdf. For more recent information, see Keegan Hamilton and Kathleen Caulderwood, "Mexican Cartels Are Arming Themselves to the Teeth With Powerful US Sniper Rifles." *Vice.* 20 August 2020, https://www.vice.com/en/article/ep48pz/how-deadly-american-sniper-rifles-became-the-mexican-cartels-favorite-weapon and Kevin Sieff and Nick Miroff, "The Sniper Rifles Flowing to Mexican Cartels Show a Decade of U.S. Failure." *Washington Post.* 19 November 2020, https://www.washingtonpost.com/graphics/2020/world/mexico-losing-control/mexico-drug-cartels-sniper-rifles-us-gun-policy/.

[10] George W. Grayson, *La Familia Drug Cartel: Implications for U.S.-Mexican Security.* Carlisle Barracks, PA: US Army War College, Strategic Studies Institute, December 2010: pp. 1-127, https://publications.armywarcollege.edu/pubs/2114.pdf and Robert J. Bunker and Alma Keshavarz, Eds., *Los Caballeros Templarios de Michoacán: Imagery, Symbolism, and Narratives.* (Small Wars Journal-El Centro eBook.) Bloomington: Xlibris, April 2019: pp.1-279, https://www.academia.edu/38806893/Los_Caballeros_Templarios_de_Michoacán_Imagery_Symbolism_and_Narratives.

[11] Robert J. Bunker, John P. Sullivan, and Alma Keshavarz, "Mexican Cartel Tactical Note #47: Anti-CJNG IAFV Trenches Dug in Michoacán." *Small Wars Journal.* 11 December 2020, https://smallwarsjournal.com/jrnl/art/mexican-cartel-tactical-note-47-anti-cjng-iafv-trenches-dug-michoacan.

For Additional Reading

Robert J. Bunker and Byron Ramirez, Eds., *Narco Armor: Improvised Armored Fighting Vehicles in Mexico.* Leavenworth, KS: Foreign Military Studies Office, US Army Command and General Staff College. October 2013: pp. 1-85.

John P. Sullivan and Robert J. Bunker, "Mexican Cartel Tactical Note #43: Improvised Armored Fighting Vehicles (IAFVs) – 'Narcotanques' and 'Monstruos Blindados' in Jalisco." *Small Wars Journal.* 10 January 2020.

Robert J. Bunker, John P. Sullivan, and Alma Keshavarz, "Mexican Cartel Tactical Note #47: Anti-CJNG IAFV Trenches Dug in Michoacán." *Small Wars Journal.* 11 December 2020.

Chapter 21

Mexican Cartel Tactical Note #49: Alleged CJNG Drone Attack in Aguililla, Michoacán Injures Two Police Officers

Robert J. Bunker and John P. Sullivan

Initially Published in Small Wars Journal on 28 April 2021

Two police officers in Aguililla, Michoacán were injured in a weaponized drone attack at approximately 0100 hours, (01:00 AM) Tuesday morning, 20 April 2021 on the highway between Aguililla and Apatzingán. The attack involving *drones artillados* (armed or 'gun' drones) is the fifth documented incident involving aerial improvised explosive devices utilized by the cartels in Mexico and the first one in which injuries have resulted.

Key Information: "Liberan policías de Michoacán bloqueo carretero entre Aguililla y Apatzingán; los atacan con drones." *Aristegui Noticias.* 20 April 2021, https://aristeguinoticias.com/2004/mexico/libera-ssp-michoacan-bloqueo-de-carretera-entre-aguililla-y-apatzingan/:

> La Secretaría de Seguridad Pública de Michoacán informó que **personal policiaco liberó el bloqueo carretero en el tramo Apaztingán-Aguililla, que**

mantenía prácticamente incomunicada de dicha población, lugar de origen de Nemesio Oceguera Cervantes. alias 'El Mencho', líder del Cártel Jalisco Nueva Generación (CJNG) y en donde este cártel mantiene una disputa por la plaza con Cárteles Unidos.

Las fuerzas policiacas retiraron un **vehículo blindado con un peso de 14 toneladas que** tuvo que ser cortado para poder ser trasladado por trozos.

El operativo estuvo encabezado por el titular de la SSP, Israel Patrón Reyes, y **comenzó desde las primeras horas de este lunes [19 April 2021]**

Medios locales reportaron que posteriormente **una célula del CJNG atacó a los policías estatales desplegados en la región con explosivos enviados en drones** en la tenencia de El Aguaje, sin que hasta el momento se conozca el alcance de estas acciones.[1]

Key Information: "El CJNG explotó un dron durante visita a Michoacán del Embajador del Vaticano; hirieron a 2 policías," *Sin Embargo.* 20 April 2021, https://www.sinembargo.mx/20-04-2021/3965563:

> **Sicarios** atacaron con un supuesto **dron** con explosivos a policías del occidental estado mexicano de **Michoacán**, previo a la visita que Franco Coppola, nuncio apostólico en México, realizará por la zona para dialogar con víctimas del narcotráfico.
>
> El Gobierno de **Michoacán** confirmó que dos agentes resultaron heridos en el ataque, ocurrido la noche del lunes en el poblado de El Aguaje, del municipio de **Aguililla**, aunque los detalles del hecho no fueron revelados.
>
> El día de hoy aprox. a las 01:00 horas, se recibe llamada de emergencia 911, reportando una agresión a personal de la Policía Michoacán, quienes se encontraban destacamentados a inmediaciones de la casa ejidal en la localidad de El Aguaje, municipio de Aguililla.

Resultando de dicha agresión 02 policías heridos, se comenta que los daños ocasionados por el ataque fueron a través de un artefacto explosivo instalado en un dron.[2]

Key Information: Jorge Monroy, "Uso de drones con explosivos, actos terroristas: SSP de Michoacán." *El Economista*. 21 April 2021, https://www.eleconomista.com.mx/politica/Uso-de-drones-con-explosivos-actos-terroristas-SSP-de-Michoacan-20210421-0092.html:

El secretario de Seguridad Pública de Michoacán, Israel Patrón Reyes, confirmó el ataque con drones a elementos de la Policía estatal en Aguililla, y dijo que acreditarse el uso de explosivos tipo C4, se trataría de un acto terrorista.

El C-4 o divergente 'Composition C-4' es una variedad común de explosivo plástico de uso bélico, y es uno de los explosivos, después del TNT, con más fuerza de los conocidos hasta el momento.

"Efectivamente, el Código Penal establece alguna conducta que refiere el uso de algunos explosivos, en este caso tenemos información de que pudiera tratarse de material explosivo conocido como C4. Considero que de aprobarse o comprobarse que están siendo utilizados estos explosivos, sin duda encuadrarían en una conducta penal que el propio Código señala como terrorismo", afirmó.

En conferencia de prensa, el funcionario de seguridad estatal también reconoció que un grupo de la delincuencia, sin mencionar al **Cártel Jalisco Nueva Generación (CJNG)** o al Cártel Unidos, tiene el control de Aguililla.[3]

Key Information: Mark Stevenson, "México: Cárteles atacan con drones cargados de explosivos." *Los Angeles Times*. 21 April 2021, https://www.latimes.com/espanol/mexico/articulo/2021-04-21/mexico-carteles-atacan-con-drones-cargados-de-explosivos:

Organizaciones del narcotráfico en México atacaron a agentes de la policía o a soldados con drones cargados de explosivos en por lo menos tres estados del país, informó el secretario de Defensa, Luis Cresencio Sandoval.

El Cártel Jalisco Nueva Generación (CJNG) fue responsable de algunos de los ataques, comentó Sandoval. Añadió que los drones cargados de explosivos han sido utilizados en los estados de Jalisco, Guanajuato y Michoacán…

…El funcionario indicó que los ataques con drones "son de preocupación", pero no han sido tan efectivos como quisieran los cárteles, debido a que los dispositivos relativamente ligeros no pueden llevar explosivos suficientes para causar daños significativos…

…Las autoridades aún no han descrito a detalle los dispositivos utilizados. Medios locales reportaron que los drones llevaban granadas de mano, pero fotografías que circulan en internet muestran que llevaban paquetes de cargas explosivas pegados con cinta adhesiva.

El ataque de esta semana ocurrió en El Aguaje, un poblado en el municipio de Aguililla, en Michoacán. El CJNG se disputa el control de ese territorio con una organización rival, la Nueva Familia Michoacana, que también es conocida como Los Viagras o Cárteles Unidos.[4]

Key Information: "Drones explosivos de Aguililla, funcionaron mal: Sedena." *MoreliActiva*. 21 April 2021, https://moreliactiva.com/drones-explosivos-de-aguililla-funcionaron-mal-sedena/:

Los drones con explosivos que fueron usados contra la Policía Michoacán en Aguililla no funcionaron correctamente, pues su capacidad destructiva fue ínfima.

Ello lo aseguró el secretario de la Defensa Nacional, Luis Crescencio Sandoval González, quien acentuó que estos artefactos únicamente pueden portar cargas pequeñas, por lo que no son capaces de dañar seriamente la vida humana.

"Los drones son de preocupación, pero no han sido efectivos, no han tenido efectividad, no pueden cargar cantidades que sean dañinas para el personal o alguna instalación, salieron dos con heridas, en el brazo y en la pierna, fue todo", acotó.[5]

Key Information: "La Sedena confirma que el CJNG ha usado drones con explosivos en Michoacán y en Guanajuato." *Sin Embargo*. 21 April 2021, https://www.sinembargo.mx/21-04-2021/3965937:

> El titular de la Secretaría de la Defensa Nacional (Sedena) confirmó esta mañana que el Cártel Jalisco Nueva Generación (CJNG) ha usado drones con explosivos, pero no sólo en Michoacán: también en Guanajuato y Jalisco. El General Luis Cresencio Sandoval González dijo que aunque les preocupa, no parecen tener demasiado impacto.[6]

Key Information: "2 to stand trial for making exploding drones in Mexico." *Mercury News*. 24 April 2021, https://www.mercurynews.com/2021/04/24/2-to-stand-trial-for-making-exploding-drones-in-mexico/:

> Prosecutors in Mexico said Saturday that two men have been ordered to stand trial for allegedly making explosive-laden drones of the kind that have been used in recent attacks on police and soldiers.
>
> The Attorney General's Office said the two were arrested a year ago in the central states of Puebla and Morelos. They face charges of violating federal firearms laws.

The evidences suggest the tactic of sending out drones with packets of explosives has been in use longer than previously thought.

Briefing Images of the Presentation and the Background Map of the 7 Road Blocks Between El Terrero and Aguililla. SSP Michoacán Briefing on Drone Attack, Tuesday, 20 April 2021. Source: SSP Michoacán

Who: The perpetrators of the attack were identified by the Government of Mexico (GoM) as members of the *Cártel Jalisco Nueva Generación* (CJNG).

What: A weaponized drone attack on police officers engaged in clearing a road from cartel obstacles near the town of Aguililla with two officers subsequently wounded.

When: Tuesday 20 April 2021 at approximately 0100 hours (01:00 AM).

Where: Apatzingán-Aguililla highway outside of Aguililla, Michoacán in El Aguaje in the municipality of Aguililla. Seven road blocks existed along this highway per the *Secretaría de Seguridad Pública del Estado de Michoacán* (SSP Michoacán) briefing, which included a map labeling the roadblocks.[7]

Why: The town is the birthplace of the CJNG leader "El Mencho" and since 2019 has been heavily contested between that cartel and *autodefensas* (self-defense forces) either allied to or controlled by *Cárteles Unidos*. The police officers were clearing roadblocks along the highway into town which resulted in their being targeted by one or more CJNG weaponized drones.

Briefing Slide 1 from SSP Michoacán Briefing on Drone
Attack, Tuesday, 20 April 2021. Source: SSP Michoacán

Briefing Slide 2 from SSP Michoacán Briefing on Drone
Attack, Tuesday, 20 April 2021. Source: SSP Michoacán

Analysis

This is the fifth documented cartel weaponized drone incident
in Mexico since October 2017. It represents a clear 'firebreak' in that
it is the first one in which injuries have resulted. The attack took
place on Tuesday 20 April at approximately 0100 hours (01:00 AM)

on Apatzingán-Aguililla highway outside of Aguililla, Michoacán. Members of the *Cártel Jalisco Nueva Generación* (CJNG)—who have engaged in weaponized drone deployments in the past—were identified as the perpetrators. The attack was directed at *Secretaría de Seguridad Pública de Michoacán* (SSP Michoacán) personnel who had been engaged in a road clearing operation into Aguililla. That town is heavily contested between the CJNG and a coalition formed between the *Cárteles Unidos* and *autodefensas*—with Mexican state and governmental level security forces intermittently engaging in stability and support operations (SASO). Two SSP Michoacán officers were lightly wounded in the drone attack.[8]

Because police—rather than civilians—were intentionally targeted in the attack, it can be viewed as an insurgent TTP (tactics, techniques, and procedures). Depending on the legal regime utilized, it can either be considered an instance of terrorism (*de facto narco* or criminal based) or a conventional force-on-force attack (potential non-international armed conflict based).

While the basics of the incident do not appear in question, a number of elements related to it are still unknown due to fragmentary, contradictory, and seemingly inaccurate news information being reported and attempts by the *Secretaría de Seguridad Pública de Michoacán* (SSP Michoacán), the Andrés Manuel López Obrador (AMLO) administration itself, and even the *Cártel Jalisco Nueva Generación*(CJNG) to shape the narrative surrounding it:[9]

- *Number of Drones and Types Involved:* The number of drones involved in the attack have been identified as two or possibly more.[10] Past cartel drone incident patterns suggest two could be plausible, however, from an attack and contextual perspective a single drone causing the injuries to the officers is far more likely. No imagery of drone components or fragments have been released by the Government of Mexico (GoM) related to this incident.[11] In fact, the release of such imagery and reports typically takes place in a haphazard and opaque manner. Still, it is relatively safe to assume that rotors drone(s) rather than fixed winged drone(s) were involved in this attack, if past incident patterns hold true.

- **Point IED Detonation or Stand-Off Bomblet Use:** Conflicting reports of the method of attack exist, with fragmentation grenades[12] or C-4 explosives[13] said to have been utilized as well as possibly IED bomblets.[14] The bomblets (pipe bomb and tailfin assembly design) said to be associated with the attack are suspect due to related sourcing, the fact they are inconsistent with typical cartel IED designs, and the weaponized drone design trajectories evident. While this is presently educated speculation—**and has not been validated**—a C-4 based IED (presumably with a shrapnel load; likely ball bearings) used for point detonation appears to make the most sense in this incident from an attack utilization and trend perspective.[15]

- **GoM Narrative:** The Mexican government is going out of its way to downplay the significance of the Aguililla incident—which it has linked to CJNG[16]—and is attempting to characterize weaponized drone attack potentials as minimal at best per the statement of the senior *Secretaría de la Defensa Nacional* (SEDENA) representative during a recent press conference with the Mexican president.[17][18][19]

- **CJNG Narrative:** The cartel has recently released a video claiming that its members were not involved in the attack and that it had instead been perpetrated by *Cárteles Unidos*.[20] The speaker states that it was that cartel, not CJNG, that created the roadblocks into the town and that they are angry at the SSP for opening the road up. As a result, they launched a drone attack against them. The speaker goes on to say that CJNG has no problem with the GoM. Note *Cárteles Unidos* (and the *autodefensas* linked/allied to them) have not been linked to weaponized drone incidents in the past, however, intelligence, surveillance, and reconnaissance (ISR) drone use has been noted.[21]

CJNG *Fuerzas Especiales* (Special Forces) Video Denying
Aguililla Drone Attack Upon SSP Michoacán. Posted
25 April 2021. Source: CJNG Social Media.[22]

- *Documented vs. Undocumented Incidents:* Questions exist
  as to whether some cartel weaponized drone incidents—such
  as seizures, interdictions, or possibly even attacks—have simply
  not been documented in publicly available news reports. New
  photos and other imagery have appeared that cannot be tied
  to documented incidents yet are said to be associated with
  the cartels. Quite possibly these simply relate to Salafi-Jihadi
  incidents taking place in the Middle East or are internet derived
  fakes and mislabeled, however, the imperative does exist for the
  GoM to suppress information related to the ongoing proliferation
  of cartel weaponized drone use as much as possible.[23]

Additionally, a few tactical elements of interest related to the incident
and its implications are as follows:

- *CJNG Night Vision Advantage Potentials:* The incident
  took place at 0100 hours (1:00 AM) under moderate-to-limited
  visibility conditions—a first quarter moon setting at 0246

hours (02:46 AM) was in effect with 50% cloud cover.[24] To date, cartel weaponized cartel drones have been assumed—given the available hardware imagery—not to have had night vision (IR; infrared) capabilities added to them.[25] However, such a capability is relatively easy to add to drones[26] and, given the past instances of cartel nighttime drug smuggling (pre-dominantly) and ISR drone use, such modifications have to be accepted to have likely taken place. Such a night vision capability would give CJNG weaponized drones an advantage over Mexican security forces who do not typically have such capabilities organic to their units—military or otherwise.

- **SSP Michoacán Counter-UAS (C-UAS) Capability:** No expectation exists that Mexican state and federal police forces would have counter-UAS capabilities or protocols in place as the cartel weaponized drone threat was, until this incident, a threat in name only. However, post-incident, this may begin to change with very basic countermeasures being implemented to small unit tactics.[27] Already, in one of the videos reviewed related to the conflict taking place in around Aguililla, a police officer can be seen standing in the back of a moving pickup truck covering the sky with his rifle presumably for C-UAS purposes as part of new force protection protocols.[28]

- **Evolving Stand-Off Bomblet Use:** At this point in time, the event still appears to have been a point detonation (i.e., kamikaze drone) type of attack—even though some reports consider it to have been based on IED bomblet(s).[29] Nevertheless, eventually crude bomblets of some sort—likely grenade (launched type) based with a tail fin assembly attachment—will start to appear. This will allow for drone reuse capabilities rather than their destruction via IED detonation in proximity of the target/ or on impact with it. Additionally, this will allow for aerial bombardment of targets from higher air-ground-levels which are harder to defeat with small arms fires. Point detonation attacks, however, will still have their own unique tactical utility for precision targeting purposes.

Conclusion

The expectation by international security analysts tracking the cartel and gang-linked criminal insurgencies in Mexico is that weaponized drone (*drones artillados*) employment will continue into the foreseeable future and will likely proliferate. The *Cártel Jalisco Nueva Generación* (CJNG) will—if past patterns hold true—most likely spearhead its use, however, *Cárteles Unidos*, *Cártel de Sinaloa* (CDS), and other cartels and their splintered factions are becoming increasingly aware of the tactical value of these systems. As a result, they will want to field their own capabilities, as has already taken place with .50 cal rifles, improvised armored fighting vehicles (IAFVs), and trenches/obstacles used to block avenues of approach into contested towns (or sections of towns) as a component of evolving *narco* urban and built-up area combat operations. Each additional documented incident is likely to present greater operational and technical sophistication. While operational security precludes complete disclosure of the technical specifications by Mexican officials, it can be expected that the cartels will continue to develop weaponized drones in pursuit of their goals.

Sources

"2 to stand trial for making exploding drones in Mexico." *Mercury News*. 24 April 2021, https://www.mercurynews.com/2021/04/24/2-to-stand-trial-for-making-exploding-drones-in-mexico/.

"El CJNG explotó un dron durante visita a Michoacán del Embajador del Vaticano; hirieron a 2 policías," *Sin Embargo*. 20 April 2021, https://www.sinembargo.mx/20-04-2021/3965563.

"CJNG 'trabaja desde el aire': utiliza drones para vigilar en Guanajuato." *La Silla Rota*. 21 April 2021, https://guanajuato.lasillarota.com/estados/cjng-trabaja-desde-el-aire-utiliza-drones-para-vigilar-en-guanajuato/509670.

"Drones explosivos de Aguililla, funcionaron mal: Sedena." *MoreliActiva*. 21 April 2021, https://moreliactiva.com/drones-explosivos-de-aguililla-funcionaron-mal-sedena/.

"Liberan policías de Michoacán bloqueo carretero entre Aguililla y Apatzingán; los atacan con drones." *Aristegui Noticias*. 20 April 2021, https://aristeguinoticias.com/2004/mexico/libera-ssp-michoacan-bloqueo-de-carretera-entre-aguililla-y-apatzingan/.

Jorge Monroy, "Uso de drones con explosivos, actos terroristas: SSP de Michoacán." *El Economista*. 21 April 2021, https://www.eleconomista.com.mx/politica/Uso-de-drones-con-explosivos-actos-terroristas-SSP-de-Michoacan-20210421-0092.html.

"La Sedena confirma que el CJNG ha usado drones con explosivos en Michoacán y en Guanajuato." *Sin Embargo*. 21 April 2021, https://www.sinembargo.mx/21-04-2021/3965937.

Mark Stevenson, "México: Cárteles atacan con drones cargados de explosivos." *Los Angeles Times*. 21 April 2021, https://www.latimes.com/espanol/mexico/articulo/2021-04-21/mexico-carteles-atacan-con-drones-cargados-de-explosivos.

Significance: *Cártel Jalisco Nueva Generación* (CJNG), *Cárteles Unidos*, Drones, *Drones Artillados* (armed or 'gun drones'), Improvised Weaponized Drones, Unmanned Aerial System (UAS)

Endnotes

[1] In English, the title reads: "Michoacán police release road blockade between Aguililla and Apatzingán; they attack them with drones." The text reads: "The Ministry of Public Security of Michoacán reported that police personnel released the road blockade on the Apaztingán-Aguililla section, which was practically isolated from said population, the place of origin of Nemesio Oceguera Cervantes, alias 'El Mencho', leader of the Jalisco Nueva Generación Cartel (CJNG) and where this cartel maintains a dispute over the square with the United Cartels."… "The police forces removed an armored vehicle weighing 14 tons that had to be cut to be able to be moved in pieces."…."The operation was headed by the head of the SSP, Israel Patron Reyes, and began from the early hours of this Monday [19 April 2021]."…."Local media reported that later a CJNG cell attacked the state police deployed in the region with explosives sent in drones in the possession of El Aguaje, without the scope of these actions being known so far."

[2] In English, the title reads: "The CJNG exploded a drone during the Vatican Ambassador's visit to Michoacán; 2 policemen were injured." The text reads: "Hit

men attacked police officers in the western Mexican state of Michoacán with an alleged drone with explosives, prior to the visit that Franco Coppola, apostolic nuncio in Mexico, will carry out in the area to talk with victims of drug trafficking.".…"The Government of Michoacán confirmed that two agents were injured in the attack, which occurred on Monday night in the town of El Aguaje, in the municipality of Aguililla, although the details of the incident were not disclosed.".…"Today at approx. 01:00 hours, an emergency call 911 was received, reporting an attack on Michoacán Police personnel, who were stationed in the vicinity of the communal house in the town of El Aguaje, in the Aguililla municipality.".…"As a result of said aggression, 2 police officers were injured, it is said that the damage caused by the attack was caused by an explosive device installed in a drone."

[3]   In English, the title reads: "Use of drones with explosives, terrorist acts: SSP of Michoacán." The text reads: "The Secretary of Public Security of Michoacán, Israel Patron Reyes, confirmed the attack with drones on elements of the State Police in Aguililla, and said that proving the use of C4 explosives would be a terrorist act.".…"The C-4 or divergent 'Composition C-4' is a common variety of plastic explosive for military use, and is one of the explosives, after TNT, with more strength than those known to date.".…"'Indeed, the Penal Code establishes some conduct that refers to the use of some explosives, in this case we have information that it could be explosive material known as C4. I believe that if these explosives are approved or proven to be used, they would undoubtedly fit into a criminal conduct that the Code itself indicates as terrorism,' he stated.".… "In a press conference, the state security official also recognized that a criminal group, without mentioning the Jalisco Nueva Generación Cartel (CJNG) or the Unidos Cartel, has control of Aguililla."

[4]   In English, the title reads: "Mexico: Cartels attack with drones loaded with explosives." The text reads: "Drug trafficking organizations in Mexico attacked police officers or soldiers with explosives-laden drones in at least three states in the country, Defense Secretary Luis Cresencio Sandoval reported.".…"The Jalisco Nueva Generación Cartel (CJNG) was responsible for some of the attacks, Sandoval said. He added that drones loaded with explosives have been used in the states of Jalisco, Guanajuato and Michoacán.".…"The official indicated that the drone strikes 'are of concern', but have not been as effective as the cartels would like, because the relatively light devices cannot carry enough explosives to cause significant damage.".…"The authorities have not yet described in detail the devices used. Local media reported that the drones carried hand grenades, but photographs circulating on the internet show that they were carrying packages of explosive charges taped together.".…"This week's attack occurred in El Aguaje, a town in the municipality of Aguililla, in Michoacán. The CJNG is fighting for control of that territory with a rival organization, the Nueva Familia Michoacana, which is also known as Los Viagras or United Cartels."

[5] In English, the title reads: "Aguililla's explosive drones malfunctioned: Sedena."
     The text reads: "The drones with explosives that were used against the Michoacán
     Police in Aguililla did not work correctly, as their destructive capacity was
     negligible." …"This was assured by the Secretary of National Defense, Luis
     Crescencio Sandoval González, who stressed that these devices can only carry
     small charges, so they are not capable of seriously damaging human life." …"'The
     drones are of concern, but they have not been effective, they have not been
     effective, they cannot carry amounts that are harmful to personnel or any facility,
     two left with injuries, in the arm and in the leg, that was all,' he said."

[6] In English, the title reads: "Sedena confirms that the CJNG has used drones
     with explosives in Michoacán and Guanajuato." The text reads: "The head of
     the Ministry of National Defense (Sedena) confirmed this morning that the
     Jalisco Nueva Generación Cartel (CJNG) has used drones with explosives, but
     not only in Michoacán: also in Guanajuato and Jalisco. General Luis Crescencio
     Sandoval González said that although they are concerned, they do not seem to
     have much of an impact."

[7] "Tapan unos, el crimen troza otros caminos, pero la Policía Michoacán seguirá
     en Aguililla, asegura Israel Patrón." *La Voz Michoacán.* 21 April 2021, https://
     www.lavozdemichoacan.com.mx/michoacan/tapan-unos-el-crimen-troza-otros-
     caminos-pero-la-policia-michoacan-seguira-en-aguililla-asegura-israel-patron/.

[8] For images, see Briefing Slide 2 from Secretaría de Seguridad Pública de
     Michoacán (SSP Michoacán) Briefing on Drone Attack, Tuesday, 20 April
     2021. Source: SSP Michoacán.

[9] See Daniel Weisz, "The Propaganda War of the CJNG and AMLO."
     Small Wars Journal. 22 April 2021, https://smallwarsjournal.com/jrnl/art/
     propaganda-war-cjng-and-amlo.

[10] The SEDENA statement said "los drones" (multiple drones) were utilized in
     the attack. Octavio Ortiz García, "Drones explosivos de Aguililla, funcionaron
     mal: Sedena." *MoreliActiva.* 21 April 2021, https://moreliactiva.com/drones-
     explosivos-de-aguililla-funcionaron-mal-sedena/. This report said two
     drones were involved in the attack. "Mexico cartel used explosive drones
     to attack police." *BBC News.* 21 April 2021, https://www.bbc.com/news/
     world-latin-america-56814501.

[11] New images from left to right: (1) What appears to be a folded DJI Mavic
     drone (background does not correspond to any known cartel weaponized drone
     incident), (2) C-4 (or similar type) explosives in plastic container with remote
     detonator on probable truck tailgate (may be linked to Tepalcatepec incident
     on 25 July 2020), and (3) Some sort of possible bomblet (does not correspond
     to any known cartel weaponized drone incident). From Briefing Slide 1 from
     the *Secretaría de Seguridad Pública de Michoacán* (SSP Michoacán) Briefing on
     Drone Attack, Tuesday, 20 April 2021. Source: SSP Michoacán.

[12] Attributed to local media reports in Michoacán. Associated Press, "Mexican drug cartels use exploding drones to attack police, soldiers." *El Paso Press*. 26 April 2021, https://www.elpasotimes.com/story/news/crime/2021/04/26/mexican-drug-cartels-use-explosive-drones-attack-police-soldiers/7384039002/. For instance, see "Con drones disparan y lanzan granadas contra policías en Aguililla, Michoacán." *Animal Politico*. 22 April 2021, https://www.animalpolitico.com/2021/04/drones-lanzan-granadas-policias-aguililla-michoacan/.

[13] Jocelyn Estrada, "Ataque contra policías en Aguililla fue con explosivos plásticos: SSP de Michoacán." *Milenio*. 21 April 2021, https://www.milenio.com/estados/ataque-policias-aguililla-explosivos-plasticos-ssp.

[14] "Niegan Los CJNG El Uso De Drones Con Explosivos Video." *Valor Por Tamaulipas*. 25 April 2021, https://www.valorportamaulipas.info/2021/04/niegan-los-cjng-el-uso-de-drones-con.html.

[15] This would be in line with the weaponized drone/IED design identified in the Tepalcatepec incident linked to CJNG taking place six-months earlier about 46 miles away from Arguililla. See Robert J. Bunker, John P. Sullivan, David A. Kuhn, and Alma Keshavarz, "Mexican Cartel Tactical Note #46: Weaponized Drones (Aerial Improvised Explosive Devices) Deployed by CJNG in Tepalcatepec, Michoacán." *Small Wars Journal*. 5 October 2020, https://smallwarsjournal.com/index.php/jrnl/art/mexican-cartel-tactical-note-46-weaponized-drones-aerial-improvised-explosive-devices. Further, the type of light wounding to the SSP officers (one in the upper right arm and the other in the lower left leg) would appear to be more consistent with a basic shrapnel ball bearing fragmentation load than larger pipe bomb fragments or, for that matter, grenade fragments (dependent on the type of grenade utilized and other considerations).

[16] Given the strategic narrative competition between the GoM and CJNG, any opportunity to further villainize that cartel (even if not technically accurate) will likely be utilized. See Daniel Weisz, "The Propaganda War of the CJNG and AMLO." *Small Wars Journal*. 22 April 2021, https://smallwarsjournal.com/jrnl/art/propaganda-war-cjng-and-amlo.

[17] Octavio Ortiz García, "Drones explosivos de Aguililla, funcionaron mal: Sedena." *MoreliActiva*. 21 April 2021, https://moreliactiva.com/drones-explosivos-de-aguililla-funcionaron-mal-sedena/.

[18] The GoM briefing video can be accessed at #ConferenciaPresidente | Miércoles 21 de abril de 2021. *YouTube*. 21 April 2021, https://www.youtube.com/watch?v=v9-b8EJhhIY.

[19] Op. Cit. See translation at note 5.

[20] Benjamin Alva, "CJNG niega ataque con drones contra Policía Michoacán." *Contramuro*. 25 April 2021, https://www.contramuro.com/cjng-niega-ataque-con-drones-contra-policia-michoacan/.

[21] Robert J. Bunker and John P. Sullivan, "Mexican Cartel Tactical Note #48: Video of CJNG Engagement of Autodefensa Mounted Infantry in IAFV in La Bocanda, Michoacán." *Small Wars Journal*. 30 December 2021, https:// smallwarsjournal.com/jrnl/art/mexican-cartel-tactical-note-48-video-cjng-engagement-autodefensa-mounted-infantry-iafv-la.

[22] For an English translation of the 2:19 minute CJNG video, see Sol Prendido, "Michoacán, Mexico: CJNG Denies Involvement in Drone Bomb Attack." *Borderland Beat*. 25 April 2021, http://www.borderlandbeat.com/2021/04/ michoacan-mexico-cjng-fuerzas.html. The contention that *Cárteles Unidos* was responsible for the attack is also discussed in "Los Drones Con Explosivos Serían De Carteles Unidos." *Valor Por Tamaulipas*. 26 April 2021, https://www. valorportamaulipas.info/2021/04/los-drones-con-explosivos-serian-de.html.

[23] As an example, the Attorney General's Office statement that the prosecution of two men linked to weaponized drone creation in Morelos and the Mexican Army statement that a weaponized drone incident took place in Jalisco have not been previously disclosed. Associated Press, "2 to stand trial for making exploding drones in Mexico." Fox 5 KVVU-TV. 24 April 2021, https://www. fox5vegas.com/news/us_world_news/2-to-stand-trial-for-making-exploding-drones-in-mexico/article_afcf0ee5-c4bc-53c9-a4e8-c0eafd61a381.html.

[24] "Moon & Sun Times Calendar for Aguililla, Michoacan (MX)" for 20 April 2021. *Solunar Forecast and Predictions*. Accessed 26 April 2021, https:// solunarforecast.com/hunting_fishing/moon_sun/calendar/mexico/michoacan/ aguililla.

[25] Still, even with such tactical advantage potentials existing, we do not know if the drone(s) involved in the attack were modified with night vision capability.

[26] See, for instance, Kalman Tihanyi, "How to Add Night Vision to Your Drones." *DroneBlog*. 3 November 2020, https://www.droneblog.com/2020/11/03/ how-to-add-night-vision-to-your-drones/.

[27] As an example of such developing protocols in the US Army, see *Counter-Unmanned Aircraft System Techniques*. ATP 3-01.81. Washington, DC: Headquarters US Army, https://rdl.train.army.mil/catalog-ws/view/100. ATSC/9B8B46D7-719C-4E15-A8FE-9F2C1E278B88-1492434973380/atp3_ 01x81.pdf.

[28] For the C-UAS/air defense imagery of the SSP police officer in the back of the pickup truck, see @adn40, "Policías asignados fueron atacados por presuntos integrantes del #CJNG en #Aguililla, dos de los policías resultaron lesionados." *Twitter*. 21 April 2021, https://twitter.com/adn40/status/1384858137358782469. Time 0:06 in the video.

[29] "Niegan Los CJNG El Uso De Drones Con Explosivos Video." *Valor Por Tamaulipas*. 25 April 2021, https://www.valorportamaulipas.info/2021/04/ niegan-los-cjng-el-uso-de-drones-con.html.

Additional Reading

David Hambling, "Mexican Cartel Injures Police Officers With Drone Bomb Attack." *Forbes*. 22 April 2021.

Robert J. Bunker, John P. Sullivan, and David A. Kuhn, "Use of Weaponized Consumer Drones in Mexican Crime War." *Counter-IED Report*. Winter 2020-2021, pp. 69-77.

Robert J. Bunker, John P. Sullivan, David A. Kuhn, and Alma Keshavarz, "Mexican Cartel Tactical Note #46: Weaponized Drones (Aerial Improvised Explosive Devices) Deployed by CJNG in Tepalcatepec, Michoacán." *Small Wars Journal*. 5 October 2020.

Chapter 22

Mexican Cartel Tactical Note #50: Additional Weaponized Consumer Drone Incidents in Michoacán and Puebla, MX

Robert J. Bunker and John P. Sullivan

Initially Published in Small Wars Journal on 10 May 2021

This research note documents two recent developments in the proliferation of weaponized consumer drones (aerial improvised explosive devices) in Mexico. The first incident is an alleged attack by the *Cártel Jalisco Nueva Generación* (CJNG) in Tepalcatepec, Michoacán on the morning of 4 May 2021. The second is the arrest of two suspected *Cártel de Santa Rosa de Lima* (CSRL) drone weaponeers in Puebla on 22 April 2021. Both incidents follow the widely reported 20 April 2021 drone attack in Aguililla, Michoacán.

Policía Michoacán (Michoacán Police) Secure a Contested Section of the Apatzingán-Aguililla Road. Source: Secretaría de Seguridad Pública de Michoacán (SSP Michoacán)

Key Information: "CJNG ataca con drones cargados de explosivos comunidad de Tepalcatepec, Michoacán." *Político MX.* 4 May 2021, https://politico.mx/minuta-politica/minuta-politica-estados/cjng-ataca-con-drones-cargados-de-explosivos-comunidad-de-tepalcatepec-michoacán/:

> **Hombres armados** presuntamente identificados como integrantes del Cártel Jalisco Nueva Generación (CJNG) atacaron con **drones cargados con explosivos** la comunidad de Pinolapa, en el municipio de **Tepalcatepec,** en el estado de Michoacán.
>
> **¿Qué se dijo?** De acuerdo con El Universal, los testimonios de los pobladores de la zona indicaron que los drones estaban cargados con explosivos C4 y granadas de fragmentación. Señalaron que el grupo armado intentó irrumpir a la comunidad cerca de las 10:00 horas.
>
> **Los hechos.** Detallaron que los sujetos armados intentaron ingresar a Pinolapa a través de la zona serrana del lugar; sin embargo, los habitantes repelieron el ataque.[1]

Key Information: "CJNG ataca con drones comunidades de Tepalcatepec, Michoacán." *El Universal*. 4 May 2021, https://www.eluniversal.com.mx/estados/cjng-ataca-con-drones-comunidades-de-tepalcatepec-michoacan:

> Sujetos fuertemente armados identificados con el **Cártel Jalisco Nueva Generación**, iniciaron esta mañana un **ataque con drones** cargados de **explosivos** en la comunidad de Pinolapa, municipio de **Tepalcatepec, Michoacán**.
>
> Los pobladores de esa zona de la **Tierra Caliente** reportaron que los artefactos no tripulados están cargados con explosivos C4 y granadas de fragmentación.
>
> Relataron que el grupo armado intentó irrumpir cerca de las 10:00 horas a esa zona limítrofe entre los municipios de Tepalcatepec y Coalcomán.[2]

Key Information: "Cártel del Tepalcatepec denunció incursión del CJNG con supuestos drones." *Infobae*. 5 May 2021, https://www.infobae.com/america/mexico/2021/05/05/cartel-del-tepalcatepec-denuncio-incursion-del-cjng-con-supuestos-drones/:

> "Se señala que en el estado de Puebla, **diversas personas se dedican a fabricar bombas utilizando un material explosivo, haciendo uso de drones para transportarlas**", indicó la FGR en un comunicado.
>
> Tepalcatepec forma parte de la región calentana que está en asedio por el CJNG que quiere recuperar la tierra que vio nacer a su líder, **Nemesio Oseguera Cervantes, el *Mencho***; mientras que remanentes locales no quieren ceder más territorio y se agruparon bajo **Cárteles Unidos**.
>
> El concilio del llamado **Cárteles Unidos** agrupa a **los Viagras, la Nueva Familia Michoacana y los Blancos de Troya**. Algunas de estas facciones **simulan ser autodefensas para legitimar retenes,**

armamento, vigilancia nocturna y rechazo de la autoridad. Sin embargo, **pretenden resistir el asedio del Cártel Jalisco Nueva Generación y no perder la plaza de Michoacán para seguir con ilícitos** como el narcomenudeo, la fabricación de droga y **extorsiones a productores de limón y aguacate.**[3]

Key Information: "Procesaron a dos fabricantes de drones con explosivos que trabajaban para el 'Marro', ex líder del CSRL." *Infobae.* 24 April 2021, https://www.infobae.com/america/mexico/2021/04/25/procesaron-a-dos-fabricantes-de-drones-con-explosivos-que-trabajaban-para-el-marro-ex-lider-del-csrl/:

> **Diego "J"** y **Rogelio "L"**, presuntos operadores del **Cártel Santa Rosa de Lima (CSRL)**, fueron **detenidos y vinculados a proceso** por su presunta participación en el **equipamiento de drones con explosivos** para las huestes de **José Antonio Yépez Ortiz, el *Marro*.**
>
> Según investigaciones de la **Fiscalía General de la República** (FGR), estos sujetos colaboraban con el CSRL desde su base en Puebla...
>
> ...Agentes de la Policía Federal Ministerial y de la Secretaría de la Defensa Nacional cumplimentaron **una orden de cateo en un domicilio de de San Andrés Cholula, Puebla,** donde fue asegurado Diego "J". De forma simultánea se desarrollaron acciones similares en la **colonia Hacienda San Antonio, de Xochitepec, Morelos,** donde cayó Rogelio "L".
>
> Un año después de recibir la denuncia anónima, el 22 de abril de 2021, los **dos supuestos operadores del CSRL fueron vinculados a proceso** por no contar con permisos para fabricar o manipular explosivos reservados a las FFAA; además de poseer armas y municiones del uso exclusivo del Ejército, Armada y Fuerza Aérea.[4]

Key Information: Jorge Butrón, "Caen en Puebla pioneros en uso de drones con explosivos." *La Razón*. 24 April 2021, https://www.razon.com. mx/estados/caen-puebla-pioneros-utilizar-drones-explosivos-432129:

> Dos fabricantes de **drones con explosivos** que trabajan para el <u>Cártel Santa Rosa de Lima</u>, fueron detenidos por violar la Ley Federal de Armas de Fuego y Explosivos y vinculados a proceso por un juez de control…
>
> …De acuerdo a las autoridades federales el 22 de marzo pasado comenzó una investigación, luego de <u>denuncias anónimas</u> que señalaron a **varias personas que producen este tipo de instrumentos** en el estado de Puebla, por ello, se concedió la autorización y orden para detener a Diego "J" y Rogelio "N"[*sic*].[5]

Who: Incident 1: *Cártel Jalisco Nueva Generación* (CJNG) Suspected.

Incident 2:*Cártel Santa Rosa de Lima* (CSRL) Suspected.

What: Incident 1: Weaponized drone attack (aerial improvised explosive device).

Incident 2: Suspected bombmakers/drone weaponeers arrested.

When: Incident 1: Tuesday 4 May 2021, at approximately1000 hours (10:00 AM)

Incident 2: Wednesday 21 April 2021, unspecified.

Where: Incident 1: Pinolapa, Tepalcatepec, Michoacán, Mexico.

Incident 2: San Andrés Cholula, Puebla, Mexico (Diego "J") and Hacienda San Antonio, de Xochitepec, Morelos, Mexico (Rogelio "L" also reported as Rogelio "N').

Why: Incident 1: Cartel competition for territorial and/or market control (CJNG).

Incident 2: Cartel competition for territorial and/or market control (CSRL).

Analysis

These two incidents represent the sixth and seventh publicly documented cartel weaponized drone incidents in Mexico since October 2017. Incident 1 is an attack on Tuesday 4 May 2021. Incident 2 involves the arrest of two suspected CSRL drone weaponeers for activity initially reported in Puebla in April 2020. These incidents follow the highly visible attack on Tuesday 20 April 2021 outside of Aguililla, Michoacán attributed to members of the *Cártel Jalisco Nueva Generación* (CJNG).[6]

The Attorney General in coordination with the Mexican Army (SEDENA) arrested Diego "J" and Rogelio "L" on Probable Federal Firearms and Explosive Charges. Source: Fiscalía General de la República (FGR), @FGRMexico, 24 April 2024.

Incident 1 involves an alleged drone attack attributed to the CJNG. It occurred during an assault on a contested road near Tepalcatepec, Michoacán. No injuries were reported.[7] Incident 2 involved the arrest of two CSRL bombmakers/weaponeers for activities in Guanajuato last year (2020) during the battle between the CSRL and CJNG.[8][9] While these two newly documented incidents are not as spectacular as the 20 April 2020 incident in Aguililla, Michoacán, they demonstrate the continuing trend toward weaponized drone proliferation among Mexican criminal armed groups (CAGs).

Presently, the CJNG is the only cartel in Mexico documented to be actively experimenting with and utilizing weaponized drones given the fragmentation of the CSRL. The expectation is that, as publicity about CJNG incidents continue, *drones artillados* use will begin to proliferate to the other cartels albeit in a haphazard manner.

Sources

Jorge Butrón, "Caen en Puebla pioneros en uso de drones con explosivos." *La Razón*. 24 April 2021, https://www.razon.com.mx/estados/caen-puebla-pioneros-utilizar-drones-explosivos-432129.

"Cártel del Tepalcatepec denunció incursión del CJNG con supuestos drones." *Infobae*. 5 May 2021, https://www.infobae.com/america/mexico/2021/05/05/cartel-del-tepalcatepec-denuncio-incursion-del-cjng-con-supuestos-drones/.

"CJNG ataca con drones cargados de explosivos comunidad de Tepalcatepec. Michoacán." *Político MX*. 4 May 2021, https://politico.mx/minuta-politica/minuta-politica-estados/cjng-ataca-con-drones-cargados-de-explosivos-comunidad-de-tepalcatepec-michoacán/.

"CJNG ataca con drones comunidades de Tepalcatepec, Michoacán." *El Universal*. 5 May 2021, https://www.eluniversal.com.mx/estados/cjng-ataca-con-drones-comunidades-de-tepalcatepec-michoacan.

"Comunicado FGR 148/21. FGR obtiene vinculación a proceso para dos hombres detenidos, uno en Puebla y otro en Morelos." Mexico City: Fiscalía General de la República (FGR). 24 April 2021, https://www.gob.mx/fgr/prensa/comunicado-fgr-148-21-fgr-obtiene-vinculacion-a-proceso-para-dos-hombres-detenidos-uno-en-puebla-y-otro-en-morelos.

"Procesaron a dos fabricantes de drones con explosivos que trabajaban para el 'Marro', ex líder del CSRL." *Infobae*. 24 April 2021, https://www.infobae.com/america/mexico/2021/04/25/procesaron-a-dos-fabricantes-de-drones-con-explosivos-que-trabajaban-para-el-marro-ex-lider-del-csrl/.

Significance: *Cártel Jalisco Nueva Generación* (CJNG), *Cártel Santa Rosa de Lima* (CSRL), Drones, *Drones Artillados* (armed or 'gun drones'), Improvised Weaponized Drones, Unmanned Aerial System (UAS)

Endnotes

[1] In English, the title reads: "CJNG attacks the community of Tepalcatepec, Michoacán with drones loaded with explosives." The text reads: "Armed men allegedly identified as members of the Jalisco New Generation Cartel (CJNG) attacked the community of Pinolapa, in the municipality of Tepalcatepec, in the state of Michoacán with drones loaded with explosives." ... **"What was said?** According to *El Universal*, the testimony of the residents of the area indicated that the drones were loaded with C4 explosives and fragmentation grenades. They indicated that the armed group tried to break into the community around 10:00 hours (10:00 AM)." ... **"The facts.** They detailed that the armed subjects tried to enter Pinolapa through the mountainous area of the place; however, the inhabitants repelled the attack."

[2] In English, the title reads: "CJNG attacks communities in Tepalcatepec, Michoacán with drones." The text reads: "Heavily armed subjects identified with the Jalisco Nueva Generación Cartel initiated an attack this morning with drones loaded with explosives in the community of Pinolapa, municipality of Tepalcatepec, Michoacán." ... "The residents of that area of Tierra Caliente reported that the unmanned devices were loaded with C4 explosives and fragmentation grenades." ... "They reported that the armed group tried to break into the border area between the municipalities of Tepalcatepec and Coalcomán at around 1000 hours [10:00 AM]." Video related to this report available at "#Video Cártel Jalisco Nueva Generación #CJNG ataca con drones comunidades de Tepalcatepec," Michoacán. *El Universal* Twitter feed (@El_Universal_Mx). 4 May 2021, https://twitter.com/El_Universal_Mx/status/1389624180241539072?s=20.

[3] In English, the title reads: "Tepalcatepec Cartel denounced CJNG incursion with alleged drones." The text reads: "'It is noted that in the state of Puebla, various people are dedicated to making bombs using an explosive material, using drones to transport them,' the FGR said in a statement." ... "Tepalcatepec is part of the Helena region that is under siege by the CJNG that wants to recover the land where its leader, Nemesio Oseguera Cervantes, El Mencho was born; while local remnants do not want to give up more territory and are grouped under United Cartels." ... "The council of the so-called United Cartels groups together the Viagras, the New Michoacán Family and the Whites of Troy. Some of these factions pretend to be self-defense groups to legitimize checkpoints,

weapons, night surveillance, and rejection of authority. However, they intend to resist the siege of the Jalisco New Generation Cartel and not lose the Michoacán Plaza [in order] to continue with illicit activities such as drug dealing, drug manufacturing, and extortion of lemon and avocado producers."

[4] In English, the title reads: "They prosecuted two manufacturers of drones with explosives who worked for 'el Marro,' former leader of the CSRL." The text reads: "Diego 'J' and Rogelio 'L,' alleged operators of the Santa Rosa de Lima Cartel (CSRL), were detained and linked to proceedings for their alleged participation in equipping drones with explosives for the hosts of José Antonio Yépez Ortiz, el Marro." ... "According to investigations by the Attorney General's Office (FGR), these subjects collaborated with the CSRL from its base in Puebla." ... "Agents of the Federal Ministerial Police and the Secretariat of National Defense completed a search warrant at a home in San Andrés Cholula, Puebla, where Diego 'J' was seized. Simultaneously, similar actions took place in the Hacienda San Antonio neighborhood of Xochitepec, Morelos, where Rogelio 'L' fell [was arrested]." ... "One year after receiving the anonymous complaint, on 22 April 22 2021, the two alleged CSRL operators were linked to the process for not having permits to manufacture or manipulate explosives reserved for the armed forces; in addition to possessing weapons and ammunition for the exclusive use of the Army, Navy and Air Force." Rogelio "L" was also reported as Rogelio "N" in some reports.

[5] In English, the title reads: "Pioneers in the use of drones with explosives fall [are arrested] in Puebla." The text reads: "Two manufacturers of drones with explosives that work for the Santa Rosa de Lima Cartel, were arrested for violating the Federal Law on Firearms and Explosives and linked to the process by a control judge." ... "According to federal authorities, on 22 March, an investigation began, after anonymous complaints indicated that several people who produce this type of instrument were in the state of Puebla, therefore, the authorization and order to arrest Diego 'J' and Rogelio 'N' was granted." ... "The arrests were made by elements of the Federal Ministerial Police, ministerial personnel, and elements of the Mexican Army at a home in the municipality of Cholula in Puebla."

[6] See Robert J. Bunker and John P. Sullivan, "Mexican Cartel Tactical Note #49: Alleged CJNG Drone Attack in Aguililla, Michoacán Injures Two Police Officers." *Small Wars Journal.* 28 April 2021, https://smallwarsjournal.com/jrnl/art/mexican-cartel-tactical-note-49-alleged-cjng-drone-attack-aguililla-michoacan-injures-two.

[7] Hiroto Saito, "Weaponized Drones in Mexico: Game-Changer or Gimmick?" *InSight Crime.* 6 May 2021, https://insightcrime.org/news/weaponized-drones-mexico-game-changer-gimmick/.

[8] See "Comunicado FGR 148/21. FGR obtiene vinculación a proceso para dos

hombres detenidos, uno en Puebla y otro en Morelos [Press Release FGR 148/21. FGR obtains connection to the process for two detained men, one in Puebla and the other in Morelos]." Mexico City: Fiscalía General de la República (FGR). 24 April 2021, https://www.gob.mx/fgr/prensa/comunicado-fgr-148-21-fgr-obtiene-vinculacion-a-proceso-para-dos-hombres-detenidos-uno-en-puebla-y-otro-en-morelos.

[9] See for example, John P. Sullivan and Robert J. Bunker, "Mexican Cartel Strategic Note No. 27: Confronting the State—Explosive Artifacts, Threats, Huachicoleros, and Cartel Competition in Guanajuato, MX." *Small Wars Journal*. 14 March 2021, https://smallwarsjournal.com/index.php/jrnl/art/mexican-cartel-strategic-note-no-27-confronting-state-explosive-artifacts-threats and Nathan P. Jones and John P. Sullivan, "Huachicoleros: Criminal Cartels, Fuel Theft, and Violence in Mexico." *Journal of Strategic Security*. Vol. 12, No. 4, 2019: pp. 1-24: https://doi.org/10.5038/1944-0472.12.4.1742.

Additional Reading

Robert J. Bunker and John P. Sullivan, "Mexican Cartel Tactical Note #49: Alleged CJNG Drone Attack in Aguililla, Michoacán Injures Two Police Officers." *Small Wars Journal*, 28 April 2021.

Robert J. Bunker, John P. Sullivan, and David A. Kuhn, "Use of Weaponized Consumer Drones in Mexican Crime War." *Counter-IED Report*. Winter 2020-2021: pp. 69-77.

Robert J. Bunker, John P. Sullivan, David A. Kuhn, and Alma Keshavarz, "Mexican Cartel Tactical Note #46: Weaponized Drones (Aerial Improvised Explosive Devices) Deployed by CJNG in Tepalcatepec, Michoacán." *Small Wars Journal*. 5 October 2020.

Conclusion

Cartel Drone Utilization Combat Trends

Robert J. Bunker and John P. Sullivan

Los Angeles, California

October 2021

The increasing threat posed by weaponized drones to US military forces and the homeland itself is well recognized by our federal officials. In April 2021, the commander of US Central Command (CENTCOM)—Marine Gen. Kenneth McKenzie—stated that "The smaller drone is a problem, and smaller drone is the future of warfare, and we need to get ahead of that right now" in reference to its ability to elude contemporary US air defense systems and threaten our foreign deployed forces.[1] While the General was specifically focusing on the threat of drones dropping explosives, he meant it within the context of Salafist-Jihadist terrorists such as the Islamic State, whose remnant forces still operate within Iraq and Syria.[2] The US Department of Homeland Security (DHS) for years now has also shared its concerns related to the threat posed by small (commercial) weaponized drones. In a November 2017 threat bulletin, foreign terrorist use of "unmanned aerial systems" was specifically mentioned as a threat to the homeland.[3] As an outcome, DHS is actively stressing Counter-UAS (C-UAS) technology development to protect our country's people and infrastructure[4] and is now deploying that technology along our Southern border.[5]

Such weaponized drone threat perceptions—and in country incidents taking place at an increasing, yet still haphazard, rate since October 2017—are by no means lost on Government of Mexico (GoM) officials. The Mexican military and other agencies have begun to quietly acquire and deploy their own C-UAS systems for force and key facility protection purposes since Fall 2020.[6] The March 2021 news report of the President of Mexico, Andrés Manuel López Obrador (AMLO), defending the deployment of anti-drone technology to protect the Presidential Palace during an International Women's Day protest from the generation of 'negative publicity' is a case in point.[7] While this is an interesting story and may signify some of the heavy-handed tendencies of López Obrador, the Presidential Palace C-UAS capability specifically exists for other reasons which are not inclusive of women and their families marching in the streets 'against rape, femicide, and other gender crimes.'[8]

CJNG Foot Soldier Launching a Recon Drone, 15 November 2020
Source: *Cártel de Jalisco Nueva Generación* (Social Media).[9]

Still, rather than a foreign Salafist-Jihadist derived terrorists threatening the Mexican homeland (which the US is concerned about), this threat specifically originates—as detailed in the twenty-two chapters of this anthology—in the ongoing criminal insurgencies being waged by its indigenous cartels and gangs. The multi-decade narco conflict is metastasizing throughout the country, which is overtly witnessing armament races between the various belligerent organized criminal organizations. While the development, acquisition, and fielding of cartel combat drones is representative of just one of these armament races, this technology is also being integrated for combined arms operations and will, at the same time, also likely impact the trajectory of other cartel armaments like IAFV design and tactics.[10] With these perceptions in mind, we will now explore three broad combat trends taking place concerning Mexican cartel drone utilization along with a emergent fourth one:

Single Use (Point Detonation) to Multi-Use (Stand Off; Bombardment) Weaponized Drones

The initial combat trend is cartel weaponized drone utilization. It is documented across seven incidents and has been chronicled within this anthology within a number of its chapters (see Annex I of this Conclusion for an incidents listing). The primary chapters discussing these incidents are Chapter 9 related to Incident No. 1 (Valtierrilla, 20 October 2017), Chapter 10 related to Incident No. 2 (Tecate, 10 July 2018), Chapter 15 related to Incident No. 3 (Puebla, 25-26 April 2020), Chapter 18 related to Incident No. 4 (Tepalcatepec, 25 July 2020), Chapter 21 related to Incident No. 5 (Aguililla, 20 April 2021), and Chapter 22 related to Incident No. 6 (linked to San Andrés Cholula, 22 April 2021) and Incident No. 7 (Tepalcatepec, 4 May 2021). These incidents have been primarily perpetrated by the *Cártel de Jalisco Nueva Generación* (CJNG) with limited *Cártel de Santa Rosa de Lima* (CSRL) involvement also documented.

These seven incidents, linked to the use of easy to fly commercial quadcopters, have been characterized as employing or relating to point detonation IEDs for single use drone missions. Improvised Explosive Device (IED) types include C4 and shrapnel, mortar shell fireworks, fragmentation grenades, and *'papa bombas.'* Still, some controversy exists if possibly Incident No. 5 (Aguililla, 20 April 2021) could be linked to one or more multi-use mission drones engaging in aerial bombardment. Early news reports related to this capability appeared after the incident on 25 April 2021 in *Brietbart* news[11] and a few days earlier in *Novedades Alcapulco.*[12] They were unsubstantiated, however, due to inconsistencies with the bomblet imagery provided (the bomblet design was not contextually correct) and cartel-linking issues.

CJNG Munitions Seized by Mexican Army Personnel belonging to the 15ᵗʰ Military Zone.
Source: *Ejército Mexicano* (Mexican Army/
SEDENA), 9 September 2021.

**Close in View of the Drone IED Bomblets (Rows
1 & 3) and Mortar Shells (Row 2).**
Source: Aseguran armamento en Mazamitla, Jalisco." *Comunicado
FGR /DPE/1930/2021.* Fiscalía General de la República
(FGR), 10 September 2021, http://ow.ly/P8ep30rTCxh;"
Jalisco (@FGR_Jal). "Aseguran armamento en #Mazamitla,
#Jalisco." *Twitter.* 10 September 2021, https://twitter.
com/FGR_Jal/status/1436421521048756228.

Early reports of CJNG aerial bombardment capabilities being
developed could not be fully substantiated until more recent imagery
had appeared. The initial imagery from 9 September 2021 pertains
to CJNG vehicular and equipment seizures by the Mexican Army
(SEDENA) and National Guard (*Guardia Nacional*) personnel in the
vicinity of Mazamitla, Jalisco.[13] In the above seizure image sourced
to Mexican Army personnel belonging to the 15th Military Zone, Rows
1 and 3 (Left Side of the Image) clearly contain IED bomblets that
have been forensically identified by David Kuhn—a standoff weaponry
expert who has been intimately involved in Mexican cartel weaponized
drone analysis.[14] Follow on video imagery posted on 22 September
2021 shows the aerial bombing (purported to be by CJNG) of a residence
in the Loma Blanca area of Tepalcatepec, Michoacán. The overhead
video shows the detonation of an IED on the roof top of a building

that has been targeted by a drone flying overhead. It is unknown if the building is associated with the local *autodefensas*, an opposing cartel, or possibly even governmental or law enforcement officials.[15]

Drone Video Screen Shots of a Targeted Building in Loma Blanca area of Tepalcatepec and Roof Detonation Plume Phases of the Dropped Aerial IED Bomblet. Video Posted 22 September 2021.
Source: *Cártel de Jalisco Nueva Generación* (Social Media).

For archival and future research purposes, additional imagery related to early (I&W) IED bomblet design and peripheral hardware potentials is being provided (See Annex II in the Conclusion). The first image pertains to recovered munitions from an engagement in early (Appx 2 or prior) March 2021 between a Mexican cartel (presumably CJNG) and SEDENA and the *Guardia Nacional* in Peribán, Michoacán, which is northwest of Tepalcatepac. The munitions were left behind in a red pickup truck abandoned by the cartel *sicarios* who initiated the ambush of the governmental forces. Seven IEDs with fins are clearly seen in an image released by SSP Michoacán. The tail finned IED bomblets appear to be derived from fragmentation grenades with a zip tie around each one securing a metal ring [presumably from which a release mechanism attached to a drone holds the bomblet in place prior to it being dropped—the same conceptual design identified with Islamic State bomblets].

The second image dates from 23 May 2021 and is from a seizure of Mexican cartel (unknown; though quite likely CJNG) hardware in Tangancícuaro, Michoacán, northwest of Peribán. The SSP Michoacán released imagery includes that of a semi-automatic pistol, pistol rounds, small plastic bags of narcotics (presumably methamphetamine), and most significantly a drone release mechanism (a DJI Mavic harness) which would be secured to a metal ring that allows narcotics or other items (such as IED bomblets) to be dropped. The third image is from cell phone video taken in a field or a parking lot under darkness (evening or early morning) behind a vehicle—likely a pickup truck or SUV—and posted to cartel social media. The cartel members (presumably from CJNG based on media reports) are manipulating a drone and an IED bomblet as they appear to begin attaching it to the drone while joking around and throwing out obscenities. The IED bomblet is artisan machine produced with a chain loop for securing to a drone release mechanism. Such an experimental design is time intensive to produce and has to be considered a transitional approach which would likely be abandoned for more easily produced bomblets (as a more institutionalized variant) based on grenade body and glued on tail fin assemblies as recovered in September 2021 by SEDENA/FGR.

The fourth image is from another CJNG social media and was posted in early June 2021. It is screen shot from a video of an eight or nine man kill team, all but one of whom are sitting in the back of a tactical assault (interior armored) vehicle. They are all decked out in 9/11 SWAT and related urban combat gear (with possible powered and/or night vision scopes). As it pertains to drone and counter-drone (or C-UAS) technology fielding, the image is significant because it clearly shows commercial radio frequency (RF) jammers attached to a number of the CJNG tactical team members. Such RF jammers have been utilized by CJNG in past raids[16] and, if configured appropriately, would provide the basis (conceptual if not actual) from which to organically provide their specialized units with a basic form of C-UAS against commercial drones. The fifth image is from yet another cartel video posted to social media. It is of a purported sicario belonging to the *Tigre Cartel de Sinaloa en Durango*, posted 15 June 2021. The sicario is dancing in the video while wearing a virtual reality (VR) rig—likely the Oculus Rift system. While a fully kitted out cartel tactical team member possibly 'rifting to an immersive music video' would at first appear harmless, this VR hardware would not be carried by such kill team personnel for relaxation purposes. Rather, we must speculate (and that is all we can do at this point) that the VR rig is likely being utilized for ISR drone C^2 purposes.[17]

Video for Propaganda, Narrative, and PSYOPS Purposes

The second combat trend is drone video use for propaganda and related purposes. The cartels have been utilizing still imagery and videos to convey their narratives to one another, the populace of Mexico, and the Federal government for decades now in their social media postings. This messaging can take on both patronage themes—such as that tied to social banditry and benevolent *jefe* (plaza boss) archetypes[18]—as well as those related to conflict and barbarism, including what can be considered the application of narcoterrorism tactics and ongoing psychological operations (PSYOPS) campaigns.[19] Drone video use by the cartels for propaganda purposes is a relatively recent event with its currently only focusing on social welfare provision.[20] Such drone video use has taken place in Tecalitlán, Jalisco in April 2020 related to the provision of COVID-19 food aid to local villagers by CJNG (Chapter 13).

We are also seeing Tepalcatepec, Michoacán area *autodefensas* (community defense forces)/*policía comunitaria* (community police) attempting to get their narratives out in their conflict with CJNG, as seen with the video of 'El Chino Drones' and cartel associates posted on social media in July 2020 (Chapter 16) and the fortified cartel building (*La Casa Balaeda*) imagery that has been released in December 2020 (Chapter 20). Earlier, the same group posted video from the aftermath of a defensive action against a CJNG intrusion, with shot up cartel vehicles and dead strewn along a stretch of road in late August 2019.[21] A much later video, originating with a 22 February 2021 social media post, provides drone video imagery from a few days earlier of a CJNG IAFV commando engagement against defending *Carteles Unidos* units.

It is related to the ongoing Tepalcatepec and La Bocanda conflict which includes local *autodefensas* (community defense forces)/*policía comunitaria* (community police) either allied to or working for *Carteles Unidos*.[22]

Drone Video Screen Shot of CJNG Commando Unit that had been Repulsed by Tepalcatepec Self Defense Forces, Late August 2019.
Source: *Policía Comunitaria Tepalcatepec* (Social Media).

Drone Video Screen Shot of CJNG Commando Unit Engagement with a *Cárteles Unidos* Defensive Position Near Tepalcatepec, 20 February 2021. Note: The drone video feed can be utilized initially for real time ISR purposes and later for propaganda purposes.
Source: *Carteles Unidos* (Social Media).

The most recent, and chilling, incident of drone-based cartel psychological operations (PSYOPS) is the daytime CJNG videoing (alleged) of the aerial bombardment of a building in Loma Blanca area of Tepalcatepec, Michoacán, posted on 22 September 2021.[23] This video represents a 'firebreak' event in that it uses a drone video feed for propaganda purposes related to an actual CJNG directed attack. This may be a component of ongoing PSYOPS with the incident being the first one to be posted in cartel social media—"Families in Aguililla have been reporting bomb-strapped drones flying over their homes since early this year [2021], in a new tactic used by the cartels to fight for their turf."[24][25] Nevertheless, the expectation is that such combat—and/or combat aftermath—cartel drone produced videos (as well as those produced by *autodefensas*) will increasingly become more common over time, as we have witnessed with the heavy Islamic State utilization of such media primarily during the 2014 through 2017 era.

ISR Integrated into Combined Arms (Mounted Infantry) Operations

The third combat trend is drone ISR integration into combined arms operations focusing on mounted infantry forces deployed in IAFV based commando units.[26] Initially, these IAFVs were deployed in ones and twos by the cartels [especially back in days of *Los Zetas* operations for *Cártel del Golfo* (CDG) in the late 1990s to about 2009] but have since been formed into much larger mounted infantry units, with CJNG social media now dominating much of the imagery related to them. These commando units and the TTPs they utilize (as well as the counter-TTPs utilized by forces defending against them) are seeing the inclusion of anti-tank weapons (typically RPGs), more .50 cal. anti-materiel rifles and machine guns (of various calibers) increasingly in armored turrets, anti-IAFV ditches which are at times linked to complex defensive positions with interlocking fields of fire, the sporadic appearance of indirect fires (mortars) and even the possibility of the use of anti-tank mines (e.g. buried pressure-detonated IEDs and/or culvert bombs).[27] We are now, over the last year, beginning to witness

ISR drones linked directly to cartel (as well as *autodefensas*) IAFVs and commando units as they evolve and the value of route and target reconnaissance missions becomes ever more recognized:

Video Screen Shot of ISR Drone Being Recovered by *Autodefensas* in the Open Air Back Compartment of an IAFV Near Bocanda, Michoacán.
Source: Marco A. Colonel (@marcolonel). *Twitter*. 17 December 2020, https://twitter.com/marcocoronel/status/1339787765513396226 (Used by Permission).

Video Screen Shot of Mounted Infantry Forces of El Señor del Sombrero (aka Mayo Zambada) with ISR Drone in Zacatecas, Posted 7 March 2021.
Source: *Cártel de Sinaloa* (Social Media).[28]

A CJNG improvised armored vehicle (IAFV) supported by a DJI
Mavic 2 Zoom aerial drone (sUAV) near Anguilla, Michoacán.
The IAFV has seen combat, as evidenced by the bullet holes in
the hood, and contains a turreted .50 cal. M2 Browning machine
gun (or foreign copy). The IAFV crew are members of the Fuerzas
Especiales Mencho (FEM), the CJNG special forces unit.
Source: "Toma Aquillia CJNG," 4 July 2021
© Cuartoscuro, used under license.

The increasing utilization of ISR by the cartels signifies the
importance planning and preparation is impacting on their operations
and the need for coordinated C^2 as offensive and defensive engagements
become more complex in both time and space. Cartel 'combat operations'
in parts of Mexico now very much resemble platoon and company size
mounted infantry actions that would be conducted by conventional
state armies. Further, weaponized drones will likely be utilized as an
additional combined arms component, as hybrid operations continue to
evolve for CJNG, *Cártel de Sinaloa* (CDS), *Cárteles Unidos* (CU), and/
or opposing *autodefensas* mobile forces.

Night Vision Combat Capability

Additionally, a fourth combat trend derived from drone night vision technology is now emerging.

Cartel narcotics smuggling and recon drones have operated under nighttime conditions for many years now, utilizing thermal cameras (FLIR; forward looking infrared) for navigation/visual flight and ISR purposes. Such low visibility use includes a narcotics bundle drop near San Luis, Arizona in November 2015 (Chapter 4), another drop near the San Ysidro Port of Entry (San Diego, California) in August 2017 (Chapter 8), counter surveillance (human trafficking 'look out') use at the Antelope Wells Port of Entry and also at Sunland Park, New Mexico in April 2019 (Chapter 12), narcotics drops in Yuma, Arizona in May and November 2020 (Chapter 14 and Chapter 19), and a weaponized drone attack directed against police officers in Aguililla, Michoacán in April 2021 (Chapter 21).

As noted by Brenda Fiegel in July 2017, "In terms of current use, drones used to transport drugs usually operate during the night, and never even land on U.S. soil. They simply drop the shipment and return to Mexico."[29] Hence, it can be readily seen that the Mexican cartels have for years now become accustomed to utilizing drone night vision capabilities, initially in their narcotics smuggling operations for navigation/visual flight purposes and later for ISR purposes. With the 0100 hours (01:00 AM) April 2021 Aguililla incident perpetrated by *Cártel de Jalisco Nueva Generación*, a 'combat firebreak' has been crossed by the cartels with their utilization of a weaponized drone's night vision capability against a targeted force—in this instance the *Secretaría de Seguridad Pública de Michoacán* (SSP Michoacán).[30]

We have scanned Mexican cartel military hardware seizures and cartel social media and video postings for well over a decade now as a component of our *SWJ-El Centro* activities and have almost never witnessed night vision technologies—such as scopes and googles—appearing in the photographic or video imagery of ground forces or other cartel personnel.[31] However, it is known from at least May 2011 via Drug Enforcement Administration (DEA) agents that night-vision goggles have been utilized by cartel lookouts from mountain tops for ISR purposes and to coordinate border smuggling operations.[32] Further, is more recently documented that Carlos and Ismael Almada, located in Guadalajara, Jalisco, purchased night-vision equipment via eBay somewhere between June through September 2019 (if not earlier in 2018) and resold it to CJNG.[33] While historical incidents related to cartel night vision combat deployment exist, the most well-known being those of the Arellano Felix Organization (AFO) in Tijuana in the 1990s and *Los Zetas* in the late 1990s and into the 2000s; they do appear to be linked to the current CJNG quest for night vision combat capability—especially as it relates to weaponized drone use.[34]

The concern over CJNG—and other contemporary cartels— deploying night vision drone technology for combat targeting purposes is that this capability would be far superior to that fielded by virtually all Mexican security and military forces except those organically possessed by a few small and elite Army (SEDENA) or Navy (SEMAR) units. This represents a component of the superiority of commercial-off-the-shelf (COTS) and its increasing battlefield impact. Additionally, for a number of years now, CJNG commando and kill team units have been appearing in cartel produced videos portraying their evolution with the inclusion of more costly and sophisticated equipment and military hardware. As has been recognized, ISR drones with night vision capabilities have recently been added to the combined arms component of IAFV commando units. This raises the potential, should CJNG and other Mexican cartels become accustomed to the tactical and operational benefits provided by drone night vision capabilities, whether such technology will not begin to proliferate into the IAFVs and onto the cartel foot soldiers themselves. From a jury-rigged perspective, removing a FLIR camera from a drone or just purchasing one online and placing it externally on a IAFV with the video viewing screen inside the crew compartment exists as future cartel night combat developmental potentialities. This would be a relatively minor technological advancement given the identified use of traditional video camera feeds into the driver's compartments of some of the more advanced CJNG IAFVs now being fielded.[35]

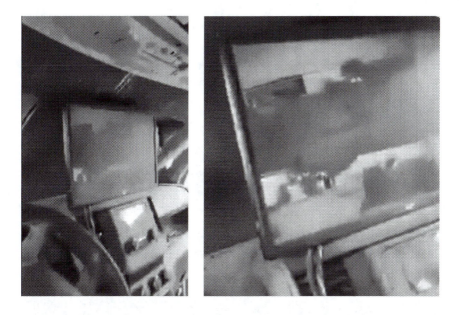

**Video Screen Shots of the Driver's Compartment Interior of a CJNG
IAFV Captured by *Cárteles Unidos* Near Tepalcatepec, 20 February
2021. Note–the Viewing Screen Has Multiple Camera Feeds.**
Source: *Cárteles Unidos* (Social Media).

Concluding Remarks

In addition to the three broad combat trends taking place concerning
Mexican cartel drone utilization—multi-use (standoff; bombardment)
weaponized; propaganda, narrative, and PSYOPs video; and ISR for
integrated mounted infantry operations—and the fourth emergent one
derived from providing night vision combat capability, a few additional
weaponized drone considerations exist. The first is the haphazard
proliferation of these drones to organized crime groups and gangs in
Latin America and the Caribbean. The appendices of the anthology
highlight weaponized drone attacks in Venezuela against President
Nicolás Manduro (Appendix 2) in August 2018 and the discovery of
weaponized drones in Colombia belonging to a FARC splinter group
in September 2019 (Appendix 3 & 4).

Additionally, grenade dropping drones were said to be a component of the Haitian President's assassination operation conducted by foreign mercenaries (per the findings of a Haitian National Police Report) which took place in July 2021 [36] and the targeting of a gang member block (presumably by an opposing gang) at Ecuadorian prison in El Litoral which was aerial bombarded by multiple drones in September 2021.[37] Of note in the latter incident is that many Ecuadorian prisons are now outfitted with C-UAS jamming devices to deny gangs the ability to drop drugs into prison yards—the El Litoral aerial bombing incident may signify that drone C^2 counter counter-measures are now being instituted by criminal gangs.[38] If this is the case, then an additional concern may exist regarding weaponized drone use spreading within various regions of the Americas.

Drug Trafficking Drone used to Smuggle Hashish and Marijuana from Morocco into Spain for Later Distribution in France
Source: *Cuerpo Nacional de Policía* (Spanish
National Police), 13 July 2021.[39]

The second is future potentials concerning the use of larger, greater flight endurance, and higher payload carrying drones beyond the smaller commercial ones (such as the DJI products) being presently utilized by the cartels in Mexico. Such drones—like the one recently seized in the port city of Málaga, Spain in July 2015—would provide the cartels with a quantum capability jump if they began to appear in their arsenals. Racks of bomblets could be attached to such drones as well as the inclusion of air-to-air and air-to-ground missiles/rockets which would provide the cartels with standoff anti-helicopter and anti-vehicular capabilities against Mexican federal forces. While this may sound speculative and alarmist, such air-to-air and air-to-ground drone-based weaponry is becoming more common out of former Soviet Union territories and other regions (using Chinese and Turkish system exports) where combat drones have been evolving and their use spreading for many years now.

The continued proliferation of improvised aerial drones by non-state actors: criminals and criminal armed groups (CAGs) can be expected as UAS becomes more common in he future. Law enforcement and military forces will be faced with these threats and need to develop countermeasures and integrate these countermeasures into their legal framework, doctrine, and training for neutralizing aerial drone threats. In addition, these security services need to continue to monitor these trends to detect future technological and tactical innovations.

Annex I: Cartel Weaponized Drone Incidents in Mexico

Number; Date	Cartel	Location	Weaponized Drone(s)	Incident
No. 1; 20 October 2017	CJNG	Near Valtierrilla, Guanajuato	1x 3DR Solo; 1x IED *'Papa Bomba'*	High-Risk Vehicle Stop Conducted by Federal Police
No. 2; 10 July 2018	CJNG (CTNG Affiliate)	Tecate, Baja	1x Tarot Ironman, 1x M-67 fragmentation grenade (or a foreign copy), 1x Chilean MK-2; 1x ISR Drone (Unknown Model)	Direct Attack Against Mexican Official for Threat Messaging Purposes (Live Grenades/Safety Pins in Place)
No. 3; 25-26 April 2020	CSRL	San Andrés Cholula, Puebla	2x DJI Inspire, 1x DJI Phantom 2; 5x 4-inch professional fireworks mortar shells; C4 Reported	Raid by Mexican Federal Authorities on CSRL Safe House
No. 4; 25 July 2020	CJNG	Tepalcatepec, Michoacán	2xMavic 2 Zoom; 4x IED Payloads (C4 & Ball Bearings)	Aborted Attack Against *Carteles Unidos*
No. 5; 20 April 2021	CJNG	Aguililla, Michoacán	1 or 2 Unidentified Rotors Drones (Probably 1); IED Likely Attached to Drone (IED Bomblet Unvalidated)	Attack on Law Enforcement Officers (SSP Michoacán) clearing a Cartel Roadblock
No. 6; 22 April 2021	CSRL	San Andrés Cholula, Puebla [Linked to No. 3]	Same Drones as No. 3	Arrest of Two Drone Weaponeers
No. 7; 4 May 2021	CJNG	Tepalcatepec, Michoacán	Unidentified IED Drone(s)	Assault Against Pinolapa Community Members (& Possibly Opposing Cartel) Defending Route into the Village

Table 1. Cartel Weaponized Drone Incidents in Mexico [40]

Annex II: IED Bomblet Variations

Image 1: Seven Tail Finned IED Bomblets which Appear Derived from Fragmentation Grenades with a Zip Tie around Each One Securing a Metal Ring.
Source: SSP Michoacán (@MICHOACANSSP).
Twitter. 2 March 2021, https://twitter.com/
michoacanssp/status/1366744645074616330.[41]

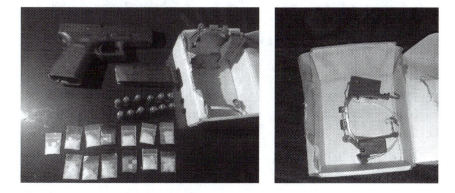

Image 2 & 2A: Left Side; Drone Release Mechanism (DJI Mavic harness).
Source: SSP Michoacán (@MICHOACANSSP).
Twitter. 23 May 2021, https://twitter.com/
michoacanssp/status/1396505495100395530.

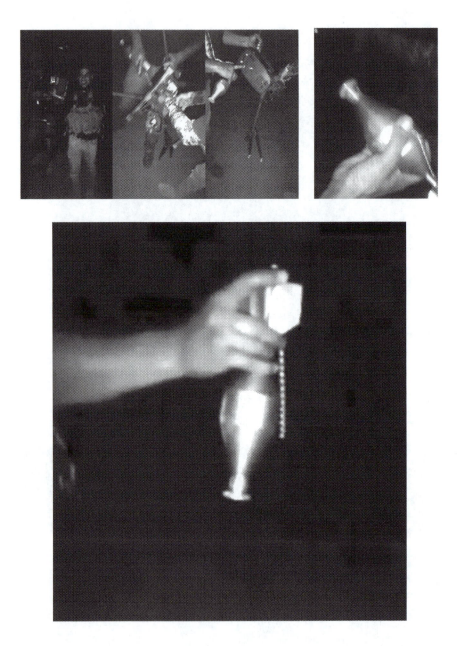

**Images 3 & 3A: Drone and Machined IED Bomblet with Chain
Attachment for Securing to the Drone.
Note Pressure Detonator at the Tip of the Bomblet.**
Source: Video Screen Shots from CJNG
(Presumed) (Social Media).[42]

**Image 4: CJNG Kill Team Members with Attached
Commercial RF Jammers, Early June 2021.**
Source: Video Screen Shot from CJNG (Social Media).[43]

Image 5: Screen Shot from a 38 Second Video of Sicarios del Tigre Cartel de Sinaloa en Durango.
Source: AG (@Jalisceinese1c). *Twitter.* 15 June 2021, https://twitter.com/Jalisciense1c/status/1404884838641934336 (Account Suspended).

Endnotes

[1] Bridget Johnson, "CENTCOM Commander: Drones Dropping Explosives 'Probably Concerns Me the Most.'" *Homeland Security Today.* 26 April 2021, https://www.hstoday.us/subject-matter-areas/airport-aviation-securit... commander-drones-dropping-explosives-probably-concerns-me-the-most/.

[2] The recognition of the threat posed by non-state violent actors (VNSAs) and criminally armed groups (CAGs) utilizing weaponized drones began to gain the

public and political spotlight in 2015 due to 'wake up incidents' taking place in London, Washington, DC, and Paris. For early perceptions of this threat, see Robert J. Bunker, *Terrorist and Insurgent Unmanned Aerial Vehicles: Use, Potentials, and Military Implications.* Carlisle Barracks: US Army War College, Strategic Studies Institute, 2015: pp. 1-74, https://apps.dtic.mil/sti/citations/ ADA623134.

[3] "Summary of Terrorism Threat to the U.S. Homeland." *National Terrorism Advisory System Bulletin.* 9 November 2017, https://www.dhs.gov/sites/default/ files/ntas/alerts/17_1109_NTAS_Bulletin.pdf.

[4] "Feature Article: As Drone Popularity and Potential Risk Soars, so too does S&T Preparedness." Science and Technology. *US Department of Homeland Security.* 16 July 2021, https://www.dhs.gov/science-and-technology/news/2021/07/16/ feature-article-drone-popularity-and-potential-risk-soars-st-prepares.

[5] Anti-drone technology—presumably to counter cartel smuggling and ISR drones (not weaponized ones)—is also beginning to be deployed along the US-Mexico border. See Aaron Boyd, "US to Deploy Anti-Drone Defenses Along US-Mexico Border." *Defense One.* 30 September 2019, https://www.defenseone. com/technology/2019/09/us-deploy-anti-drone-defenses-along-us-mexico-border/160247/ and Ken Klippenstein and Alex Emmons, "Border Police Want a Bite of the Burgeoning Anti-Drone Industry." *The Intercept.* 3 May 2021, https://theintercept.com/2021/05/03/cbp-border-drones-military/.

[6] Noé Cruz Serrano, "Mexico will purchase anti-drone systems to fight drug cartels." *El Universal.* 23 September 2020, https://www.eluniversal.com. mx/english/mexico-will-purchase-anti-drone-systems-fight-drug-cartels. In Spanish, see Noé Cruz Serrano, "Ejército va por sistema antidrones de cárteles del narco." *El Universal.* 21 September 2020, https://www.eluniversal.com.mx/ nacion/ejercito-va-por-sistema-antidrones-de-carteles-del-narco.

[7] "AMLO defends measures to protect palace, points to possibility of drone attack." *Mexico News Daily.* 10 March 2021, https://mexiconewsdaily.com/news/ amlo-defends-measures-to-protect-palace-points-to-possibility-of-drone-attack/.

[8] "20,000 march in Mexico City for justice for women and an end to the violence." *Mexico News Daily.* 9 March 2021, https://mexiconewsdaily.com/ news/20000-march-in-mexico-city-for-justice-for-women/.

[9] CJNG social media republished as "FOTO: Así son los drones con explosivos que el CJNG usa para atacar a rivales y hasta autoridades." *La Opinion.* 19 July 2021, https://laopinion.com/2021/07/19/foto-asi-son-los-drones-con-explosivos-que-el-cjng-usa-para-atacar-a-rivales-y-hasta-autoridades/ and "#Sicarios del #CJNG se graban utilizando drones en la sierra de #Michoacán, alparecer se comenta que estos drones seran usados con explosivos c4 en contra del gobierno que se meta apoyar a #LosViagras #MexicoRojo #EU." *Blog Del Narco Oficial.* 15 November 2020, https://www.facebook.com/BlogDelNarcoOficialMx.

[10] For up-to-date information and analysis of Mexican cartel armaments, see the sister publication, Robert J. Bunker and John P. Sullivan, Eds., *Illicit Tactical Progress: Mexican Cartel Tactical Notes 2013-2020*. (A Small Wars Journal–El Centro Anthology.) Bloomington: Xlibris, 2020.

[11] "Exclusive Photos: Mexican Cartels Weaponize Drones to Drop IEDs." *Brietbart*. 25 April 2021, https://www.breitbart.com/border/2021/04/25/exclusive-photos-cartels-in-mexico-weaponized-drones-to-drop-ieds/.

[12] "Atacan con drones a policías en Michoacán." *Novedades Acapulco*. 20 April 2021, https://novedadesaca.mx/atacan-con-drones-a-policias-en-michoacan/. Also tweeted at https://twitter.com/novedadesaca1/status/1384668865783689216 on 20 April 2021.

[13] "Aseguran Ejército y Guardia Nacional un arsenal en Mazamitla." Agencia Quadratín. 9 September 2021, https://colima.quadratin.com.mx/jalisco/aseguran-ejercito-y-guardia-nacional-un-arsenal-en-mazamitla/ and Carolina Solis, "Armamento y 6 autos, algunos con blindaje artesanal, son decomisados en Mazamitla, Jalisco." *Debate*. 9 September 2021, https://www.debate.com.mx/policiacas/Armamento-y-6-autos-algunos-c...artesanal-son-decomisados-en-Mazamitla-Jalisco-20210909-0109.html.

[14] Robert J. Bunker email correspondence with David A. Kuhn between 11-15 September 2021 related to CJNG aerial bomblets forensic analysis.

[15] Ciro Gómez Leyva (@CiroGomezL), Tweet; "Esta es la grabación de un supuesto drone del #CJNG usado para lanzar explosivos. En las imágenes se ve cómo suelta una carga y explota sobre una casa en la zona conocida como Loma Blanca, en #Tepalcatepec, #Michoacán:" See 1:03 minute video of the aerial bombardment and news commentary. *Twitter*. 22 September 2021, https://twitter.com/cirogomezl/status/1440883348356964353?s=21.

[16] Robert J. Bunker and Alma Keshavarz, "Signal Inhibitors Seized in Two Recent Raids Against CJNG in Mexico." *OE Watch*. Vol 10. Issue 11. December 2020: p. 75.

[17] Robert J. Bunker, "Virtual Martyrs: Jihadists, Oculus Rift, and IED Drones. TRENDS Research & Advisory." 14 December 2014, http://trendsinstitution.org/virtual-martyrs-jihadists-oculus-rift-and-ied-drones/ [Original article removed]. Reprinted as Essay 1 in Robert J. Bunker, *Terrorism Futures: Evolving Technology and TTPs Use*. (A C/O Futures Pocketbook.) Bloomington: Xlibris, 2020.

[18] See, for instance, John P. Sullivan and Robert J. Bunker, Editors, *Covid-19, Gangs, and Conflict*. (A Small Wars Journal–El Centro Reader.) Bloomington: Xlibris, 2020.

[19] See Robert J. Bunker, Alma Keshavarz, and John P. Sullivan, "Mexican Cartel Tactical Note #39: GoPro Video Social Media Posting of Cártel Santa Rosa de Lima (CSRL) Tactical Action against Cártel Jalisco Nueva Generación (CJNG)

in Guanajuato – Indications & Warning (I&W) Concerns." *Small Wars Journal.* 5 March 2019, https://smallwarsjournal.com/jrnl/art/mexican-cartel-tactical-note-39-gopro-video-social-media-posting-cartel-santa-rosa-de-lima.

[20] Interestingly enough, no overhead drone videos showing off cartel—specifically CJNG—military hardware and firepower, such as that of one of its commando units, has been released to date.

[21] "Video: Brave vigilantes arm themselves to protect their town from Mexico's Jalisco New Generation Cartel." *Daily Mail.* 4 September 2019, https://www.dailymail.co.uk/video/news/video-1999176/Video-Drone-shows-aftermath-shooting-cartel-vigilantes.html and "Drone footage of the Abuelo vs Mencho battle yesterday in Tepalcatepec, Michoacan." *Borderland Beat.* 31 August 2019, http://www.borderlandbeat.com/2019/08/drone-footage-of-abuelo-vs-mencho.html.

[22] Antonio Aranda (@antonioaranda_), "Imágenes del enfrentamiento entre sicarios de Cárteles Unidos celebran la derrota del #CJNG en Michoacán municipio de Tepalcatepec." Twitter. 24 February 2021, https://twitter.com/antonioaranda_/status/1364792451991444141 and "Talcatepec, Michoacán: Drone Footage of CJNG Failed Incursion." *Borderland Beat.* 25 February 2021, http://www.borderlandbeat.com/2021/02/tepalcatepec-michoacan-drone-footage-of.html.

[23] Ciro Gómez Leyva (@CiroGomezL), Tweet; "Esta es la grabación de un supuesto drone del #CJNG usado para lanzar explosivos. En las imágenes se ve cómo suelta una carga y explota sobre una casa en la zona conocida como Loma Blanca, en #Tepalcatepec, #Michoacán:" See 1:03 minute video of the aerial bombardment and news commentary. *Twitter.* 22 September 2021, https://twitter.com/cirogomezl/status/1440883348356964353?s=21.

[24] Luis Chaparro, "'Like a flying ant': An operative describes how Mexico's cartels use drones to attack enemies and smuggle drugs." *Business Insider.* 1 June 2021, https://www.businessinsider.com/how-mexicos-cartels-are-using-drugs-for-attacks-drug-smuggling-2021-5.

[25] An additional CJNG PSYOPS campaign may be the threatened or actual inclusion of carbofuran (a toxic chemical) to its IED bomblets. See "Con drones, CJNG lanza explosivos mezclados con insecticida mortal contra el Ejército en Michoacán." *YouTube.* 4 October 2021, https://www.youtube.com/watch?v=PZ0cA2PRl4g&t=1s.

[26] A basic cartel IAFV primer is Robert J. Bunker and Byron Ramirez, Eds., *Narco Armor: Improvised Armor Fighting Vehicles in Mexico.* Fort Leavenworth: US Army Foreign Military Studies Office (FMSO), 29 October 2013, https://community.apan.org/wg/tradoc-g2/fmso/m/fmso-monographs/197127.

[27] Concerning the possible use of an improvised anti-tank mines, see "VIDEO: Explota con mina camión blindado del CJNG, sicarios del Mencho así cayeron en trampa." *La Opinión.* 3 January 2021, https://laopinion.com/2021/01/03/

video-explota-con-mina-camion-blindado-del-cjng-sicarios-del-mencho-asi-cayeron-en-trampa/.

[28] "Zacatecas, Mexico: Cartel de Sinaloa Announce Their Presence Online." *Borderland Beat.* 7 March 2021, http://www.borderlandbeat.com/2021/03/zacatecas-mexico-cartel-de-sinaloa.html. The original CDS Facebook account containing these videos has either been removed, deleted or its security settings have been changed. It was at https://www.facebook.com/Zacatecas-es-del-Sr-Del-sombrero-143324887593089/videos/zacatecas/189998215857954/.

[29] Brenda Fiegel, "Narco-Drones: A New way to Transport Drugs." *Small Wars Journal.* 5 July 2017, https://smallwarsjournal.com/jrnl/art/narco-drones-a-new-way-to-transport-drugs.

[30] This acquisition of this combat capability, along with the use of a virtual reality headset (VR rig), was earlier expressed regarding Salafist-Jihadists. See Robert J. Bunker, "Virtual Martyrs: Jihadists, Oculus Rift, and IED Drones." *TRENDS Research & Advisory.* Terrorism Futures Series. 14 December 2014, http://trendsinstitution.org/?p=762.

[31] This is not to say such technology is not being utilized in some instances by cartel personnel in infrared red sniper scopes and other low light optics—we are just not seeing it (or have been able to reliably identify it) in open-source imagery releases and cartel social media.

[32] Mark Potter, "Cartels using Ariz. mountaintops to spy on cops." *NBC News.* 19 May 2011, https://www.nbcnews.com/id/wbna43096521. See also Stephen Dinan, "Drug scouts for cartels would be sent to prison under proposed bill." *Washington Times.* 27 March 2017, https://www.washingtontimes.com/news/2017/mar/27/drug-scouts-for-cartels-would-be-sent-to-prison-un/.

[33] Nathaniel Janowitz, "How Mexico's Most Powerful Cartel Used EBay to Arm Themselves With Military Gear." *Vice.* 6 April 2021, https://www.vice.com/en/article/epdj77/how-mexicos-most-powerful-cartel-used-ebay-to-arm-themselves-with-military-gear?utm_source=reddit.com.

[34] See Douglas Farah, "Cartels Hire Mercenaries to Train Security Forces." *Washington Post.* 4 November 1997, https://www.washingtonpost.com/archive/politics/1997/11/04/cartel-hires-mercenaries-to-train-security-forces/faeb4801-a082-4f53-941c-49bd8cfae93c/ and Daniel Epstein and ProPublica, "How DEA Agents Took Down Mexico's Most Vicious Drug Cartel." *The Atlantic.* January-February 2016, https://www.theatlantic.com/magazine/archive/2016/01/drug-enforcement-agency-mexico-drug-cartel/419100/. The expectation is Los Zetas cartel members—given their Mexican special forces backgrounds—would have also employed night-vision googles with their kill team personnel. This was cited by Lisa J. Campbell, "Los Zetas: operational assessment." *Small Wars & Insurgencies.* 12 March 2010: pp. 55-80. Other

later cases of night vision equipment purchases for the cartels exist such as those taking place from January 2012 through March 2013 by Jaime Jauregui in Orlando, Florida. See James Varney, "Would-be Mexican drug cartel arms merchant sentenced in New Orleans." *New Orleans Advocate.* 19 March 2015 (Updated 17 July 2019), https://www.nola.com/news/crime_police/article_f6802307-75e5-5c8c-b107-525163257656.html.

[35] Antonio Aranda (@antonioaranda_), "Sicarios de Carteles Unidos celebran la derrota del #CJNG en Michoacán mancillando el cadáver de un sicario, reportan la muerte del 'M2' líder del #CJNG en el Estado. Decir que se comportan como hienas es ofender a los animales." *Twitter.* 24 February 2021, https://twitter.com/antonioaranda_/status/1363985280194150400.

[36] Jacqueline Charles and Jay Weaver, "Grenade-dropping drones, a paranoid president, guards who ran: Latest on Haiti assassination." *Miami Herald.* 20 September 2021 (Updated), https://www.miamiherald.com/news/nation-world/world/americas/article254275213.html.

[37] David Hambling, "Drug Cartels Carry Out Drone Bombings, Evade Jammers." *Forbes.* 1 October 2021, https://www.forbes.com/sites/davidhambling/2021/10/01/drug-cartels-carry-out-drone-bombings-evade-jammers/?sh=2988489c3ecc and "Reportan ataque con drones a cárcel en Guayaquil, Ecuador." teleSURtv. 13 September 2021, https://www.telesurtv.net/news/ecuador-reportan-ataque-drones-carcel-guayaquil-20210913-0016.html.

[38] David Hambling, "Drug Cartels Carry Out Drone Bombings, Evade Jammers."

[39] *Policía Nacional* (@policia), "Intervenido en Málaga un dron de 4,35 metros preparado para el trasporte de droga entre #Marruecos y #España." *Twitter.* 13 July 2021, https://twitter.com/policia/status/1414875097752653826?s=20. English translation: A 4.35 meter drone was seized in Málaga, prepared to transport drugs between #Morocco and #Spain; It belonged to an organization dedicated to buying hashish and marijuana in Morocco for subsequent transport and distribution in France; 4 people were arrested.

[40] An earlier version of this Table with four incidents identified was published in Robert J. Bunker, John P. Sullivan, and David A. Kuhn. "Use of Weaponized Consumer Drones in Mexican Crime War." *Counter-IED Report.* Winter 2020-21, pp. 69-77, https://www.academia.edu/45000611/Use_of_Weaponized_Consumer_Drones_in_Mexican_Crime_War.

[41] See also "Balacera en Peribán: emboscan sicarios a oficiales de la ley." *Primeraplana.* 2 March 2021, https://primeraplana.mx/archivos/798417.

[42] For publicly available reposting access; (Left & Middle Images) Charbell Lucio, "Dos elementos de la GN lesionados, tras enfrentamiento con CJNG en Chinicuila." *Revolucion3.0.* 25 May 2021, https://revolucion.news/dos-elementos-la-gn-lesionados-tras-enfrentamiento-cjng-chinicuila; (Right Image & Original

Video) christopher (@just_some_d00d), "Another video with much clearer look at the explosive." *Twitter.* 27 May 2021, https://twitter.com/just_some_d00d/status/1397935567367778310; (Bottom Image) christopher (@just_some_d00d), "And there it is! These things look professionally machined." *Twitter.* 25 May 2021, https://twitter.com/just_some_d00d/status/1397383142013652995/photo/1.

[43] A publicly available reposting of the video can be accessed at CMF Global Risk (@CMF_GlobalRisk). "CJNG (Jalisco New Generation Cartel) welcomes its new sponsor @511Tactical?" *Twitter.* 9 June 2021, https://twitter.com/CMF_GlobalRisk/status/1402770084318285829. The original cartel posting may have been from https://twitter.com/codigorojocdmx/status/1402665516179804167 but the account is now suspended.

Afterword

Narco Drone Futures

Conrad 'Andy' Dreby and Scott Crino

Washington, DC

September 2021

Our company, Red Six Solutions, collaboration with Robert Bunker and John Sullivan goes back quite a while. We first met and began working with Robert eight years ago when he participated in a forward-looking tabletop exercise we ran for the Department of the Army. The threat environment for that exercise was Central America and one of the applied technologies we considered were drones. Eight years seems like yesterday, but in the story of small drones it is a millennium. It was just at the time SZ DJI Technology Co., Ltd. of China or DJI released their DJI Phantom 3 with DJI's first live-streaming action camera. That combination along with Wi-Fi and smartphone technology made DJI's Phantom 3 the first-to-market integrated drone platform at the consumer level. Almost immediately, as DJI sales took-off, we began to see drones being used widely for illicit purposes.

After our first meeting, our professional relationship with Dr. Bunker continued and was cemented in 2015 when he joined us to assess counter unmanned aerial systems (UAS) for a project with the

Central Intelligence Agency (CIA). The culminating event of the project was a live-simulation in which we flew small UAS against various sUAS detection systems positioned at the CIA headquarters in Langley, Virginia. It was a just a 2-day event but it was formative for Red Six because it was our first chance to run a drone exercise in the highly restrictive Washington, DC airspace. And, to our knowledge, it was the first time any private company had ever done so.

After the CIA event, providing drone services became a major business line for us. The demand for our services came in part due to the perspective which we bring which is from the viewpoint of the enemy. Our company's motto is *Sempre Inimicus*, Latin for always the enemy. At Red Six, we believe the best way for clients to learn about their strengths and weaknesses is to contest their ideas intellectually and their defenses physically using realistic threats simulations. In order to deliver this type of service for UAS, we knew we would have to study threat actors closely. Unfortunately, after we started, we quickly learned there are no extant sources of regularized, open source UAS incident reporting. There were research projects and studies, and for some major events, major media reporting, but the day in and day out collection and analysis of drone incidents did not exist so we began to do the work ourselves. Soon we found there was a market for this information and so, for the last three years, we have produced monthly UAS incident reports covering the use of drones by irregular militaries, terrorists, and criminal organizations.

Our information gathering and analysis on UAS incidents brings us in regular contact with Drs. Bunker and Sullivan. With Robert and John, we compare notes with their providing critical context on narco cartels and their operating environments while we lend our knowledge of drone technologies and piloting. It is a good partnership. An example of our working together was in our analysis of the May 2020 incident in Puebla, Mexico, where armed drones were discovered Mexican federal authorities in San Andrés Cholula. We used information Robert and John provided in our monthly report to clients and our analysis on the drones' capabilities contributed to what became Robert and John's Mexican Cartel Tactical Note #45 in *Small Wars Journal*.

With regard to narco cartels, Red Six has every reason to believe we will continue to work regularly with Robert and John for the foreseeable future. It is our belief that drones are a more integral part of cartel operations than is generally understood and that their use will continue to expand over time. Drone use by cartels is very much under reported. If a DJI Phantom overloaded with methamphetamine crashes in San Ysidro, California, it is reported locally. Or when US Customs and Border Protection (CBP) makes a press release on seizing drones from a safehouse in San Luis, Arizona, it makes the news. But, in general, the everyday employment of drones for reconnaissance and surveillance is missed. Even the occasional incidents which do make the news are easily forgotten because there is no regularized accounting in the public sphere of their having happened. It is Red Six's assessment that narco cartels currently use drones in their operations thousands of times every month. Currently a systematic approach for tracking suspected drug cartel drone flights along the US border with Mexico does not exist. However, based on our firsthand observations and anecdotal information, Red Six assesses there are likely thousands of potential drug-cartel drone flights on the border every month.

From an operational perspective, one reason drone use by narco cartels will grow is because technological advances will continuously enhance their return on investment value. Improvements in flight endurance and load capacity, combined with ever increasing accuracy of commercial global navigation systems, will mean drug smugglers can fly large semi-autonomous drones further and further across the border. Even now, large multirotor drones can carry 10-to-12 lbs. payloads miles across the border, making it very hard for law enforcement to interdict them.

Besides carrying cargo, cartels will continue their extensive use of UAS for reconnaissance and surveillance. Small but sophisticated camera drones are inexpensive, easy to operate, and invaluable to illicit traffickers because they can be used to spot the location or absence of law enforcement. These capabilities can be especially valuable for certain tasks, such as when cartels want to move members across the border who are wanted by United States law enforcement. In those instances, small drones can provide the cartels an advantage by enabling the cartels to see and avoid any areas where they may be at risk of arrest.

While there are many other drone manufacturers, for the near future, narco cartels will continue to rely primarily on DJI products. Cartels will have this expected preference for the same reasons commercial drone buyers predominantly choose DJI products which is because of their reliability, quality of engineering, and performance. The reliance of cartels on DJI products, however, offers law enforcement some opportunities for countering cartel drone operations. As a company, DJI has cooperated with United States authorities, and when requested, has established no fly zones (NFZ) where DJI products are unable to fly; such, NFZs can be used to deny certain areas to cartel operations. DJI also offers government customers a product called AeroScope, which is a system that effectively tracks all DJI products in a covered area. This information can be used to track patterns of behavior and could be used to develop intelligence on drone operations. Moreover, because of their wide popularity, DJI products are widely studied by the growing counter UAS developer community and there now exists a host of commercial systems which can fairly effectively detect and track DJI products too. Thus, the future is not completely a one way street in favor of the cartels with regard to using DJI products.

While DJI products will likely remain a mainstay of the cartels' drone fleets, a likely reason for them to eventually move to other UAS platforms will be the need to carry larger payloads. Multirotor UAS, like the ones described so far, have limited ranges and payload capabilities and the two work against one another. That means any increase in range reduces the payload and vice versa. A way around this limitation is by using larger fixed-wing drones which can fly longer and carry more. For a long time, large fixed-wing drones have been used for smuggling in other parts of the world. In the Ukraine and Belarus, there are many cited instances of large fixed-wing drones being used to smuggle cigarettes and contraband into the European Union to avoid taxes. And just recently, a large vertical takeoff and landing UAS, which was being used to smuggle drugs into the country from North Africa, was confiscated by Spanish National Police.[1] That particular aircraft had an almost 15 ft. wingspan (4.5 meters or 14.8 feet) with seven hours of endurance, and was capable of carrying a 330 lb. (150 kilos) payload.[2] It makes sense that these aircraft, well suited for flying illicit cargo deep into the United States, will someday be used by narco cartels.

There have only been a few incidents in which the cartels are also known to use weaponized drones. This year, there were several reported incidents of the CJNG using drones armed with grenades in Mexico's Michoacán State to attack village defense forces and, in April, the CJNG used a small UAS to drop a grenade which injured two policemen. While both these incidents were relatively small in scale, they do show the willingness of cartels to use drones offensively. In other parts of the world, notably Islamist militants in Syria and Russian-backed separatists in Ukraine, drones were initially used primarily for observation platforms and it took time for the groups to develop their capabilities and tactics for employing weaponized drones. The recent incidents in North America strongly suggest the cartels may also be moving in this direction.

A challenge in forecasting the directions narco cartels will go with drones is their wealth. When studying the use of drones by militants, terrorists, and criminal organizations elsewhere in the world, there is usually an observable degree of cost consciousness seen in their aircraft designs. The affects of tight budgets are seen in the materials used to build the airframes and the selection of the components used inside the aircraft. Overseas, it is not unusual to see drones with cardboard wings and secondhand flight components. Unlike these groups, cartels have plenty of money, so they will be able to keep up with all the latest developments in drone technology and choose those which work the best for them.

The current status quo for drone use by cartels is likely to change in the near future, In the existing state, cartels use drones primarily for observation and surveillance, with some smuggling primarily in built-up areas where there is a strong CBP presence, and occasional armed drone attacks against rivals and police. Any tightening of security along the United States – Mexico border will incentivize the cartels and other drug smuggling groups to seek out alternative UAV platforms which will enable them to carry larger loads over longer distances. Limited funding has been allocated to the CBP, as well as the Coast Guard, for counter UAS systems. As the border's network of counter UAS systems matures, cartels will likely seek to procure systems which are less easy to detect than the commercial systems they are using now. As violent as the cartels are notoriously known to be, it also seems likely they will over time increase their use of weaponized drones against rival groups and law enforcement organizations. And, as is typical with violent extremist groups, the cartels will use their social media accounts to promote the attacks to attract recruits and terrorize communities. While unfortunate, this expected outcome means Red Six and Drs. Bunker and Sullivan will continue sharing our knowledge with one another for a long time to come.

Endnotes

[1] Nacho Sáchez, "Spanish police seize large drone used to carry drugs from Morocco." *El País*. 15 July 2021, https://english.elpais.com/spain/2021-07-15/spanish-police-seize-large-drone-used-to-carry-drugs-from-morocco.html.

[2] See Ameya Paleja, "Police Capture Massive 5-Engine Drone That's Used for Drug Transportation." *Interesting Engineering*. 23 August 2012, https://interestingengineering.com/spanish-police-capture-5-engine-drone-used-for-drug-transport; and Cuerpo Nacional de Policía (policianacional), "Intervenido en Málaga un dron de 4,35 metros preparado para el trasporte de estupefacientes entre #Marruecos y #España" (Intercepted in Malaga a drone of 4.35 meters prepared to transport narcotics between #Morocco and #Spain.) *Instagram*. 13 July 2021, https://www.instagram.com/p/CRQ1QWkoCjf/?utm_source=ig_web_copy_link.

Postscript

Anticipating Future Threats
of Unmanned Systems

James J. Torrence

Jerusalem, Israel

August 2021

As I reflected on what I learned after reading this anthology, I could not help but think we are at the beginning of an unmanned systems revolution. Dr. Bunker and Dr. Sullivan are always on the forefront of capturing use cases of unmanned systems by drug cartels and non-state actors. Their work also shows that law enforcement and military personnel are often surprised by the innovations and uses of unmanned systems by criminal organizations. There are myriad lessons one can learn from this book, but I want to focus on the importance of looking to the future to identify possible use cases of unmanned systems. If threat analysts only consider the past and present to develop strategies for the use of unmanned systems by cartels and criminal organizations, their strategies will always be outdated. In this post-script, I will contemplate the future with a focus on the following trends that I think will shape the impact of unmanned systems for the drug cartels, and other criminal organizations: biomimicry and animal replication, hacking self-driving cars and delivery drones, and 3D (Three Dimensional) printing with nanomaterials and explosives.

Biomimicry & Animal Replication

Every unmanned system featured in this study was clearly identifiable as a man-made electronic device. In the next decade, further advances will be made using biomimicry that will result in unmanned systems blending in with their surroundings. Biomimicry is the "conscious emulation of nature" to produce "anything from a single object to a large technological system, even a city, in an environmentally sustainable manner."[1] Some unmanned systems can already "operate in and around wild animals without disturbing them, moving in the water, air and on the ground with natural motions."[2] For aerial unmanned systems, "bird-like designs may also help winged drones fly with greater endurance and agility."[3] The shift towards biomimicry is moving rapidly and will frustrate current detection systems.

Currently, radar systems face challenges "differentiating between a small bird and a drone."[4] The unmanned systems featured in this study are (or will be) detectable by advanced millimeter-wave radar systems that detect "the spin of the drone's rotor blades."[5] But, even advanced "acoustic and radar detection will become more difficult with the increased use of biomimicry and synthetic feather flying systems, which will minimize noise and remove hard surfaces and blade rotation that are detectable by radar."[6]

The "holy grail" of biomimicry for scientific researchers is an artificial beetle "with a body of appropriate shape, mechanical and thermal properties, and color; the ability to fly; and the ability to sense and communicate."[7] The only thing holding back the creation of a beetle is the "miniaturization of batteries, sensors, motors, and transmitters" and "optical-flow sensing systems."[8] Beetles and other artificial animals will soon propagate the environment and will be used for more than research. As this anthology makes clear, nefarious actors are willing to pay a premium for technical talent that can advance the usage of unmanned systems. Between government advances and non-state actors innovating to circumvent detection, unmanned systems will continue to leverage advances in miniaturization and biomimicry.

Implications of Biomimicry & Animal Replication

Biomimicry and animal replication will create the conditions for multi-domain operations with commercial unmanned systems. Radar developers would be smart to start employing ornithologists, marine biologists, and other animal experts in the development of smart radars to prepare for a future when detecting the difference between an unmanned system and an animal will come down to subtle details that will not be accounted for in a general detection system.

Imagine unmanned systems that mimic a mischief of rats, a school of fish, a catastrophe of moles, and a flock of birds that operate autonomously, infrequently receive position, location, and information updates, communicate with each other, and blend into their natural environment. These advances in unmanned systems create opportunities to transport narcotics through underground tunnels just big enough for small rodents while simultaneously providing ISR through bird-like unmanned systems. Additionally, the unmanned systems masquerading as schools of fish can provide ISR for narcotics transport from larger manned or unmanned systems.

A criminal organization that can harness biomimicry can create the ultimate Internet of Things (IoT) which will create a more resilient network and positioning system. More unmanned systems that blend into the natural environment create the opportunity for navigation by triangulation (which can combine with inertial navigation) if the Global Positioning System (GPS) is blocked during an operation. Additionally, unmanned systems in the air, in the sea, in subterranean regions, and on land create opportunities to propagate signals and allow communication in an environment where an adversary is trying to degrade communications and electronic systems. If unmanned systems can blend in with other animals, the mobility, dispersion, and concealment created will result in uncertainty for a criminal organization's adversary and act as a deterrent against attempts to disrupt unmanned systems.[9]

Hacking Self-Driving Cars & Delivery Drones

This book highlighted examples of the development and modification of commercial drones to conduct attacks, ISR (intelligence, surveillance, and reconnaissance), and delivery of narcotics. What if cartels and other criminal organizations did not need to develop their own unmanned systems for ISR? What if the existing infrastructure for self-driving cars and delivery drones creates the conditions to tap into existing routes, deliveries, and trips for information? Self-driving cars and delivery drones are an evolution of unmanned systems that state actors and non-state actors can use to obtain information.

Self-driving cars represent a prime target for criminal organizations looking to conduct surveillance. There are predictions that "750,000 autonomous-ready cars will hit the roads in the year 2023, which means they'll be vulnerable to attack. That represents, by some estimates, more than two-thirds of cars on the road, riddled with attack surfaces both known and not."[10] Defense experts recognize that "no matter how hard we try and how complex we make the security solutions on vehicles, it is impossible to make something perfectly secure and unhackable."[11] State and non-state actors will leverage weaknesses in self-driving cars to conduct ISR (imagine a cartel listening into Drug Enforcement Administration (DEA) conversations in the back of a self-driving car on strategy) and potentially even for violence.[12]

Delivery drones will result in the ubiquitous use of unmanned systems worldwide. In Oranmore, Ireland, there is a drone delivery trial in which the "technicians run the trial, loading deliveries—that can weigh up to 2 kilograms—into a white paper bag, which is placed into a removable cargo bay inserted into the drone."[13] The drones in the trial are "set to fly at 50 mph, the drones cruise at 260 feet to reach their destination and can reach anywhere in the town within two minutes."[14] When the drones arrive at a house for a delivery, "the drone lowers to 80 feet before a hatch on its belly opens and the bag gently spirals to earth at the end of a thin rope."[15] Instead of developing their own systems, criminal organizations can tap into existing delivery drones for ISR (using the drone's own cameras and addresses) to monitor people at their homes and on the highway without ever having to deviate from the normal drone flight paths.

Implications of Hacking Self-Driving Cars & Delivery Drones

Currently, cartels and other criminal organizations are paying substantial sums of money to buy and modify commercial unmanned systems to put into operation. If cartels and other non-state actors can take advantage of ubiquitous unmanned systems to conduct ISR and obtain information, it will allow them to focus on other uses of unmanned systems. They may still need unmanned systems for ISR, but if they can listen in on car rides, watch traffic routes, and observe people's houses by tapping into self-driving cars and delivery drones, they will be able to focus more time and technical expertise on the precision use of unmanned systems (whether for ISR, narcotics delivery, or assassination).

The vulnerabilities in self-driving cars and delivery drones coupled with their ubiquity (multiple companies with multiple different software operating systems) means that criminal organizations will benefit from the information available (location, video, recordings, photographs, etc.). The focus from threat analysts is often on what innovations criminal organizations will develop to thwart detection, but the paradigm changes when they can tap into existing network infrastructure to observer what unmanned systems are doing as part of an approved legal framework.

3D Printing with Nanomaterials & Explosives

There were not any examples of 3D printed unmanned systems in this book, but that will soon change. 3D printing of both explosives and nanomaterials will fundamentally alter the threat of unmanned systems.

The examples in this book show that the weight of additional explosives created a drone that was challenging to fly and not easy to detonate. Imagine a 3D printed drone that is resistant to infrared detection, can harness solar power, and is comprised of plastic explosives allowing for a controlled or targeted detonation.

Government organizations have already started researching the 3D printing of explosives. The Defence Science and Technology Laboratory (DSTL) "an executive agency sponsored by the Ministry of Defence (MOD) of the UK, has started to develop 3D printed explosives."[16] DSTL "aims to create new possibilities for various explosive effects using intricate designs enabled by 3D printing, and also to reduce transportation and storage costs."[17] 3D printing "is disrupting many industries, and the explosives market represents just another example."[18] It is just a matter of time before the research in 3D printed explosives results in unmanned systems being printed with explosive material.

Nano-materials, black silicon specifically, are another example of how 3D printing will change the threat level of unmanned systems. Black silicon "can be used as the surface of the absorption layer to absorb near-infrared light, to enhance the photodiode absorption of light."[19] "Black silicon absorbs light" and "incoming light reflects back and forth between the vertical spires, bouncing around within the material instead of escaping."[20] Black silicon is also emerging as a nanomaterial that can be used as a solar cell.[21] Black silicon traps light making it nearly undetectable by infrared radar and its ability to harness solar energy has future implications for power and flight time.

Implications of 3D Printing Nanomaterials & Explosives

Printing unmanned systems with black silicon will immediately make them harder to detect, thus creating more opportunities (especially at night) to operate without detection. Using plastic explosives as a base material also means that the flight problems experienced by mounting explosives (e.g., grenades) will no longer apply. 3D printing also allows for different shapes and sizes, which is perfect for the gradual shift to biomimicry and animal replication. Imagine unmanned aerial systems that mimic a flock of birds printed with a base of plastic explosive and an outer layer of black silicon. The undetectable, explosive flock is just one example of how 3D printing will escalate the threat of unmanned systems.

Final Thoughts

This anthology, along with studies and edited volumes that Dr. Bunker and Dr. Sullivan plan to complete in the next few years, will be used to understand the nascent stages of the evolution of unmanned systems. Instead of being surprised by the innovative ways in which cartels and criminal organizations use unmanned systems, threat experts must look at emerging trends and think about how examples seen in this book will evolve in the future. There are many trends that will impact the use of unmanned systems, but I think biomimicry and animal replication, hacking self-driving cars and delivery drones, and 3D printing with nanomaterials and explosives are the most important.

Endnotes

[1] Michael Fisch, "The Nature of Biomimicry: Toward a Novel Technological Culture," *Science, Technology, & Human Values.* Vol. 42, no. 5, September 2017: p. 797, http://www.jstor.org/stable/26405615.

[2] Thomas G. Pledger, *The Role of Drones in Future Terrorist Attacks.* Land Warfare Paper 137. Association of the United States Army, Institute of Land Warfare; Arlington, VA, February 2021, https://www.ausa.org/sites/default/files/publications/LWP-137-The-Role-of-Drones-in-Future-Terrorist-Attacks_0.pdf.

[3] Charles Choi, "Mimicking Biology for Better Drones." *Inside Unmanned Systems.* 17 December 2020, https://insideunmannedsystems.com/mimicking-biology-for-better-drones/.

[4] Thomas G. Pledger, *The Role of Drones in Future Terrorist Attacks.* Land Warfare Paper 137.

[5] Ibid.

[6] Ibid.

[7] Michael H. Bartl and Akhlesh Lakhtakia, "The Artificial Beetle, or a Brief

Manifesto for Engineered Biomimicry." Proc. SPIE 9429, *Bioinspiration, Biomimetics, and Bioreplication* 2015, 94290B, 26 March 2015: p. 7, https://doi.org/10.1117/12.2083656.

[8] Ibid.

[9] James Torrence, "Lessons for Cyber Policymakers." *The InterAgency Journal*. Vol. 10, no. 1, 2019: pp. 50-54, https://thesimonscenter.org/wp-content/uploads/2019/02/IAJ-10-1.pdf.

[10] Stephen Ornes, "How to Hack a Self-Driving Car." *Physics World*. 18 August 2020, https://physicsworld.com/a/how-to-hack-a-self-driving-car.

[11] Charlie Miller, "Lessons Learned from Hacking a Car." *IEEE Design & Test*. Vol. 36, no. 6, December 2019: p. 7, doi: 10.1109/MDAT.2018.2863106.

[12] The Amazon TV show *Upload* features an impressive example of assassination using a self-driving car.

[13] Patrick Reevell, "Future of Drone Delivery May be Happening in This Small Irish Town." *ABC News*. 13 June 2021, https://abcnews.go.com/International/future-drone-delivery-happening-small-irish-town/story?id=78158015.

[14] Ibid.

[15] Ibid.

[16] Michael Petch, "UK Defence Agency Plans to 3D Print High Explosives." 3dprintingindustry.com. 16 March 2020, https://3dprintingindustry.com/news/uk-defence-agency-plans-to-3d-print-high-explosives-169082/.

[17] Ibid.

[18] Jeff Kerns, "A Look Inside the 'Explosive' 3D-Printing Industry." Machinedesign.com. 26 January 2018, https://www.machinedesign.com/3d-printing-cad/article/21836373/a-look-inside-the-explosive-3dprinting-industry and Homeland Security Advisory Council, "Final Report of the Emerging Technologies Committee 3D-Printing." Department of Homeland Security. 24 February 2020: p. 18, https://www.dhs.gov/sites/default/files/publications/final_report_hsac_emerging_technology_subcommittee_3dprinting_508_compliant.pdf.

[19] Zheng Fan et al., "Recent Progress of Black Silicon: From Fabrications to Applications." *Nanomaterials*. Vol. 11, no. 41, 2021: p. 14, https://doi.org/10.3390/nano11010041.

[20] Sam Million-Weaver, "'Stealth' material hides hot objects from infrared eyes." Wisconsin.edu. 21 June 2018, https://news.wisc.edu/stealth-material-hides-hot-objects-from-infrared-eyes/.

[21] Zheng Fan et al., "Recent Progress of Black Silicon: From Fabrications to Applications;" pp. 16-17.

Appendix 1

Narco-Drones in Colombia

Brenda Fiegel

Initially Published in OE Watch January-February 2017 Issue, p. 35

"... drug cartels have realized that using drones to transport drugs does not carry as much risk or cost as much as traditional means of transport (human traffickers, narco tunnels, semi-submersibles)."

Source: "Descubren un 'narcodron' en Colombia; enviaba cocaína a Panamá (Authorities Discover Narco-Drone in Colombia with Cocaine Shipment Headed to Panama)," *Excelsior*, 15 November 2016. http://www.seguridadenamerica.com.mx/ noticias/de-consulta/secciones-revist-seguridad-en-america/noticias-de-control-de-acceso/24684-descubren-un-narcodron-en-colombia-enviaba-cocaina-a-panama:

The new system [the drone] was discovered last Tuesday [15 Nov] in Colombia during an anti-drug operation near the town of Bahia Solano, in the jungle area of the Choco Department, where 130 kilos of cocaine buried in the beach were discovered along with parts of a remote control aircraft ready to be assembled, indicated General Jose Acevedo, Regional Police Commander.

Source: "Policía revela 'el hormigueo', nueva modalidad del Clan Úsuga para sacar droga del país (Police Reveal New Micro-Trafficking Scheme Utilized by Clan Úsuga to Move Drugs from Colombia to Panama)," *Noticias CM*, 20 October 2016. http://www.cmi.com.co/policia-revela-el-hormigueo-nueva-modalidad-del-clan-Úsuga-para- sacar-droga-del-pais:

> *Clan Usaga has known trafficking routes that extend from Uraba, Colombia into Darien, Panama. Recently, this group began using migrants (Cuban, African, and Asian) to transport drugs along this route. As part of this new modus operandi, migrants carry anywhere from 25-30 kilograms of cocaine through the Darien province. This journey is said to take up to 5 days.*

Source: "Narco envía droga a EU... en drones (Drug Trafficking Organizations Using Drones to Send Drugs to US)," *Ejecentral*, 17 August 2014. http://www.eluniversal. com.mx/nacion-mexico/2014/carteles-fabrican-narcodrones-trasiego-eu-1022274.html:

> *Mexican drug cartels are using drones to transport drugs across the US/Mexico border. Authorities first became aware of this phenomenon in 2011 after multiple drones detected by radar systems were shot down upon their entrance into the United States. Since their first detection, two significant changes associated with drone use by cartels have been reported. First, Mexican drug cartels are now placing orders to drone producers in Mexican-based cities including the Federal District, Guadalajara, Monterrey, Querétaro and Tijuana. This is a change because at the start of 2011, virtually all drones owned and operated by cartels were produced abroad; primarily in Israel. Second, drug cartels have realized that using drones to transport drugs does not carry as much risk or cost as much as traditional means of transport (human traffickers, narco tunnels, semi-submersibles).*

OE Watch Commentary: In mid-November [2016], Colombian Police seized a drone used by narco- traffickers to send small cocaine shipments from the Colombian jungle to Panama. This represents the first case in which authorities have detected this trafficking method in Colombia. It appears that drug cartels are discovering that using drones to transport drugs is cheaper and less risky, indicating that this will likely be an increasingly popular method. The accompanying passages from Latin American sources discuss this phenomenon.

This first accompanying passage discusses that the drone was discovered during an anti-drug operation in Colombia. Drones currently being utilized for this activity can transport 10 kilograms of cocaine and travel up to 100 kilometers in a single trip. This method was likely developed by a group known as the Clan del Golfo (formerly known as Clan Úsuga) which is the largest criminal gang in Colombia dedicated to narcotrafficking and illegal mining. As the second passage notes, this same group is also known for its diverse trafficking methods which also include using illegal migrants to transport shipments from Colombia into Panama.

The use of drones by drug cartels is a new phenomenon in Colombia; but as the third passage notes, this tactic has been used by Mexican cartels since 2011. In fact, by 2012, drone use along the US-Mexican border was highly prevalent as evidenced by the US interception of 150 drones carrying an estimated two metric tons of drugs, primarily marijuana. Today, drones are being classified as the "perfect drug mule" in Mexico and abroad as they involve less risk to drug trafficking organizations and their employees. The third passage further points out they can even be produced domestically in Mexico. Additionally, they cost significantly less than drug tunnels and semi-submersibles and are even capable of transporting cash shipments. The popularity of drone use by drug cartels in Mexico and Colombia is likely to increase; especially if producers work on devolving more agile models that can carry more weight and fly at lower altitudes. **End OE Watch Commentary (Fiegel)**

Appendix 2

Alleged Assassination Attempt on Venezuelan President Nicolás Maduro

Robert J. Bunker and Alma Keshavarz

Initially Published in OE Watch September 2018 Issue, p. 66

Security using shields to protect President Nicolás Maduro. 8/4/18.

"They were two remotely piloted aircrafts [Drone model DJIM600], each loaded with 1kg of explosives."

Source: "¿Qué se sabe del atentado contra el presidente Nicolás Maduro? (What is known about the attack against President Nicolas Maduro?)," *TeleSUR*, 5 August 2018. https://www.telesurtv.net/multimedia/maduro-claves-atentado-venezuela-20180805-0035.html:

> *...They were two remotely piloted aircrafts [Drone model DJIM600], each loaded with 1kg of explosives. Signal jammers were used to disorient the first drone that was flying around the perimeter to attack...*

Source: "Maduro asegura que involucrados en atentado en su contra se entrenaban con personal colombiano (Maduro said that those involved in the assassination attempt trained with Colombian personnel)," *El Universal*, 7 August 2018. http://www.eluniversal.com/politica/17050/maduro-asegura-que-involucrados-en-atentado-en-su-contra-se-entrenaban-con-personal-colombiano:

> *...It is clear and there is enough evidence that shows it was the Colombian administration of Juan Manuel Santos. We have their location, we have their names—in Chinácota, north of Santander, where they trained with other Colombian assassins and terrorists...*

OE Watch Commentary: On 4 August 2018 two drones equipped with explosives allegedly attempted to assassinate Venezuelan President Nicolás Maduro. According to the accompanying excerpted article from *TeleSUR*, a Caracas-based news outlet, Maduro was delivering a televised speech commemorating the 81st anniversary of the Bolivarian National Guard during a military parade in Caracas, Venezuela. Per *TeleSUR*, the drones were a model DJI M600, each armed with one kilogram of C-4. Venezuela's Communications Minister, Jorgé Rodriguez, confirmed that the attack occurred near the end of the event on Bolivar Avenue in Caracas. The drones were allegedly diverted by state military forces using radio frequency jammers, during which one drone lost control and exploded near the Don Eduardo apartment complex, damaging a portion of the building. According to the article from *El Universal*, a Caracas based newspaper, the attack wounded seven state security officials. Venezuelan Interior Minister Néstor Reverol stated that six individuals have been detained and are being questioned. President Maduro released a statement hours after the attack blaming "ultra-right Venezuelans and ultra-right Colombians" and former Colombian President Juan Manuel Santos for the attempt on his life. **End OE Watch Commentary (Keshavarz and Bunker)**

Addendum (Anthology)

One of the drones exploding in front of the stage wounding seven guards:

"President Nicolas Maduro Reveals Evidence For Drone Attack." *Telesur*. 7 August 2018, https://www.youtube.com/watch?v=TPD3WSW6wAs. [Venezuelan Government/No Restrictions on Use]

One of the drones crashing into the Don Eduardo apartment building:

"President Nicolas Maduro Reveals Evidence For Drone Attack." *Telesur.* 7 August 2018, https://www.youtube.com/watch?v=TPD3WSW6wAs. [Venezuelan Government/No Restrictions on Use]

Appendix 3

Drones Pose New Threat on Colombia's Pacific Coast

Maria Alejandra Navarrete

Initially Published in InSight Crime on 25 September 2019

The army destroyed two drones laden with explosives found in Tumaco

The discovery of two drones in the department of Nariño has raised fears about what impact such technology could have on the current conflict in the southwest of Colombia.

On September 19, the Colombian Army announced that an operation had seized[1] two Syma drones, loaded with 600 grams of explosives, on the road connecting the municipalities of Pasto and Tumaco, in the department of Nariño.

According to investigators[2] in charge of the operation, from the Special Brigade Against Drug Trafficking (Brigada Especial contra el Narcotráfico – BRACNA), two detonators and various types of shrapnel were found alongside the explosives.

According to authorities, the drones and detonators came from Ecuador and Peru.

In an official press release,[3] the army declared that the drones allegedly belonged to the Oliver Sinisterra Front (Frente Oliver Sinisterra – FOS),[4] dissidents of the Revolutionary Armed Forces of Colombia (Fuerzas Armadas Revolucionarias de Colombia – FARC). The FOS, currently led[5] by alias "Comandante Gringo," was allegedly planning to carry out attacks against the military and civilian population in Tumaco.

Signal inhibitors were used to carry out the operation, conducted in coordination with the police, so as to avoid the devices being activated remotely during the controlled destruction of the explosives.

InSight Crime Analysis

While there have been reports[6] of drones allegedly being used by criminal groups in Colombia for surveillance purposes, this is the first time drones have been registered as a weapon and their impact on the armed conflict in Nariño is unclear.

The financial and technological investment involved is not difficult, as these devices can easily be obtained on the open market. But the potential use of drones to transport explosives in a remote part of

Colombia could reflect ex-FARC Mafia groups trying out new combat strategies.

The FOS, previously led by Walter Patricio Arizala, alias "Guacho,"[7] has been known for its use of unconventional weapons, such as anti-personnel mines, fuel bombs triggered by passing army troops, and hidden stashes of explosives that detonated when army helicopters landed nearby, as *Semana*[8] reported.

However, the advantages drones could provide, or other plans that the group may have for employing the use of this technology, are yet not clear. According to information[9] divulged by authorities, the explosives were stuck to the drones to completely destroy the vehicle at the moment of detonation.

Although authorities believe they were to be deployed against public security forces, it is worth noting the FOS has been weakened of late, due to disputes[10] with the United Guerrillas of the Pacific (Guerrillas Unidas del Pacífico – GUP) and a criminal group led by alias "Contador"[11] as they vie for control of one of Colombia's largest coca cultivation enclaves.

Endnotes

[1] "Ejército halla dos drones cargados con explosivos en zona rural de Tumaco, Nariño." *El País*. 18 September 2019, https://www.elpais.com.co/judicial/ejercito-halla-dos-drones-cargados-con-explosivos-en-zona-rural-de-tumaco-narino.html.

[2] "Ejército desactivó dos drones cargados con explosivos en Tumaco." *El Tiempo*. 18 September 2019, https://www.eltiempo.com/justicia/investigacion/ejercito-desactivo-dos-drones-cargados-con-explosivos-en-tumaco-413816.

[3] "Neutralizados y destruidos dos drones cargados de explosivos." *La Seguridad es de Todos*. 19 September 2019, https://www.ejercito.mil.co/index.php?idcategoria=467681.

[4] Parker Asmann, "Battles Between Former FARC Groups Displace Hundreds in Colombia." *InSight Crime*. 11 March 2019, https://www.insightcrime.org/news/analysis/colombia-forced-displacement-farc/.

[5] "Destruyen drones con explosivos que pertenecerían a disidencias en Nariño." *BLU Radio*. 18 September 2019, https://www.bluradio.com/nacion/destruyen-drones-con-explosivos-que-pertenecerian-a-disidencias-en-narino.

[6] "Narcos, disidencias y ELN estarían utilizando drones para espiar a la fuerza pública." *BLU Radio.* 19 September 2019, https://www.bluradio.com/nacion/narcos-disidencias-y-eln-estarian-utilizando-drones-para-espiar-a-la-fuerza-publica.

[7] "Walter Patricio Arizala, alias 'Guacho.'" *InSight Crime.* 18 January 2019 (Last Update), https://www.insightcrime.org/colombia-organized-crime-news/walter-patricio-arizala-alias-guacho/.

[8] "Drones bomba: la nueva arma de las disidencias en Tumaco." *Semana.* 19 September 2019, https://www.semana.com/nacion/articulo/drones-bomba-la-nueva-arma-de-las-disidencias-en-tumaco/632441/.

[9] "Ejército desactivó dos drones cargados con explosivos en Tumaco." *El Tiempo.* (See Note 1).

[10] Parker Asmann, "Battles Between Former FARC Groups Displace Hundreds in Colombia." (See Note 4).

[11] Ibid.

Appendix 4

Neutralizados y destruidos dos drones cargados de explosivos

La Seguridad es de Todos

First Published in La Seguridad es de Todos 19 September 2019 [1]

Drones cargados de explosivos
Source: Colombian Military [No Restrictions on Use]

En las últimas horas, tropas de la Brigada contra el Narcotráfico del Ejército Nacional, en una acción coordinada con la Policía Nacional, logran la ubicación y neutralización controlada de dos drones Syma cargados de explosivos.

Los drones fueron ubicados en la vereda Alto Agua Clara, municipio de Tumaco, en Nariño, gracias al trabajo de inteligencia militar del Batallón Contra el Narcotráfico N.º4 que desarrolla labores de apoyo operacional en ese sector del país. Allí, en un área boscosa y escondidos se encontraban los drones que estaban cargados con 600 gramos de explosivo plástico, dos cordones de seguridad o llamada «mecha lenta», dos detonadores y material tipo metralla (clavos, tuercas y tornillos).

Así, en coordinación con la Policía Nacional, se logra la destrucción controlada de los drones que pertenecerían presuntamente al grupo armado organizado residual Óliver Sinisterra, el cual pretendería atentar contra las tropas que garantizan la seguridad y defensa del municipio de Tumaco y el departamento de Nariño, así como contra la población civil. Esto demuestra cómo estos grupos recurren a la utilización de medios y métodos que infringen los derechos humanos y el derecho internacional humanitario.

El material hallado en los drones quedó a disposición de la autoridad competente para determinar el tipo de explosivo que sería utilizado y continuar el proceso de judicialización pertinente. Con este resultado se logra neutralizar y actuar oportunamente ante posibles métodos de terrorismo usados por los grupos armados organizados, ratificando una vez más el compromiso del Ejército Nacional y las Fuerzas Militares en estar siempre atentas ante hechos de criminalidad y terrorismo que afectan la seguridad y defensa de la Nación.

Endnotes

[1]   https://www.ejercito.mil.co/index.php?idcategoria=467681.

Addendum (Anthology) Translation

Two explosive-laden drones neutralized and destroyed
Security Belongs to Everyone
Photo caption: Drones loaded with explosives

In the last few hours, troops of the National Army's Anti-Narcotics Brigade, in a coordinated action with the National Police, managed to locate and neutralize two Syma drones loaded with explosives.

The drones were located in the village of Alto Agua Clara, municipality of Tumaco, in Nariño, thanks to the military intelligence work of the Anti-Narcotics Battalion No. 4, which carries out operational support tasks in that sector of the country. There, in a wooded and hidden area, the drones were loaded with 600 grams of plastic explosive, two safety cords or "slow fuse," two detonators and shrapnel-type material (nails, nuts and bolts).

Thus, in coordination with the National Police, the controlled destruction of the drones that allegedly belonged to the residual organized armed group Óliver Sinisterra, which intended to attack the troops that guarantee the security and defense of the municipality of Tumaco and the department of Nariño, as well as the civilian population, was achieved. This demonstrates how these groups resort to the use of means and methods that violate human rights and international humanitarian law.

The material found in the drones was handed over to the competent authority to determine the type of explosive that would be used and to continue the relevant judicial process.

With this result, it is possible to neutralize and act in a timely manner in the face of possible terrorist methods used by organized armed groups, ratifying once again the commitment of the National Army and the Military Forces to always be alert to criminal and terrorist acts that affect the security and defense of the Nation.

Appendix 5

The Proliferation of Drone Use by Drug Traffickers

Brenda Fiegel

Initially Published in OE Watch February 2020 Issue, p. 72

OE Watch Commentary: Buen Jarin, Peru is inhabited by an indigenous community called the Tikuna. This community is located along Peru's triple border, which is shared by Colombia and Brazil along the Amazon River. As the first passage from the Peruvian news source El Confidencial discusses, areas around the community are infiltrated by drug traffickers dedicated to producing cocaine and cutting down valuable trees utilized to make high end products for export. To thwart this activity and to monitor the community, an NGO called the Rainforest Foundation has provided drones. According to indigous leader Pablo García Akawasa, the drone donation is useful because deforestation is so extensive in Buen Jarin that their land is "turning into a pampa (extensive treeless plains)."

Drone use is not exclusive to Peru, or to NGOs or community leaders. Colombian and Mexican cartels also utilize this new technology as discussed in El Confidencial. José Acevedo, a Commander of the Regional Police in Colombia stated that, "Drones are utilized by criminal groups such as the National Liberation Army (ELN) and the Revolutionary Armed Forces of Colombia (FARC) to transport cocaine to Panama. These drones can carry up to 10 kilograms of cocaine." Furthermore, drone use is not exclusive to drug trafficking as evidenced by the seizure of one in September 2019. The drone seized in Colombia is believed to have belonged by the FARC and contained a plastic explosive device filled with nails, screws, and nuts. Had the explosive detonated, it would have caused extensive damage.

As for drones in Mexico, cartels have been using them since 2010, but have recently began using these devices to monitor authorities, carry out attacks with explosives, and monitor illegal migrants crossing the border with the help of coyotes as this represents a secondary business of many criminal organizations in Mexico. The passage from Zeta Tijuana discusses the use of explosives in Mexico. **End OE Watch Commentary (Fiegel)**

Source: "Del narco al 'coyote': el mundo del crimen latinoamericano reinventa el uso del dron (From Drug Trafficking to Coyote: Latin America's Criminal World Reinvents Drone Use)," *El Confidencial*, 1 January 2020. https://www.elconfidencial.com/mundo/2020-01-01/del-.

> According to Pablo García Akawasa, a leader within the indigenous Tikuna community, the drone donation from the Rainforest Foundations is useful as these drones can detect new areas of deforestation as well as locations in which drug traffickers are operating.
>
> ...
>
> Mark Cancian of the Center for Strategic and International Studies (CSIS) indicated that drone use in not exclusive to Peru's indigenous community. He further indicated that drug traffickers in both Colombia and Mexico have access to this technology and that "drones are now accessible to many groups not associated with NGO's to include criminal organizations." Furthermore, he indicated that "the drones being utilized are not as

sophisticated as those used by the United States, but instead are purchased commercially and then modified to carry light drug shipments, messages, drugs, mobile phones, and even explosives." Drones are also used to collect intelligence regarding positions and activities of local authorities and to monitor drug shipments and drug trafficking routes. Finally, he indicated that drones utilized to carry drugs have the capacity to move shipments of 5 kilos or more and can travel up to 100 miles.

Source: "Con drones envían granadas a casa de Sosa Olachea (Armored Drones Used to Attack Residence of Sosa Olachea)," *Zeta Tijuana*, 10 July 2018. https://zetatijuana.com/2018/07/con-drones-envian-granadas-a-casa-de-sosa-olachea.

Mexican cartels are once again changing the rules of drug trafficking by armoring their drones to conduct criminal attacks. The first reported attack was carried out In July of 2018 when an unnamed drug carted utilized an armed drone was used to attack the house of Baja California State Public Safety Secretary Gerardo Sosa Olachea in the city of Tecate along the US-Mexico border. At least two drones were allegedly used in the strike. The first was armed with audio and video equipment and two improvised explosive devices (IEDs) that did not explode after falling into the official's yard. A second drone was seen above the house likely performing surveillance.

Selected Bibliography

A

Benjamin Alva, "CJNG niega ataque con drones contra Policía Michoacán." *Contramuro*. 25 April 2021, https://www.contramuro.com/cjng-niega-ataque-con-drones-contra-policia-michoacan/.

"La amenaza." *Zeta*. 16 July 2018, http://zetatijuana.com/2018/07/la-amenaza-3/.

"Aseguran en Puebla drones y explosivos para realizar actos terroristas." *Telefono Rojo*. 3 May 2020, https://telefonorojo.mx/aseguran-en-puebla-drones-y-explosivos-para-realizar-actos-terroristas/.

Associated Press, "2 to stand trial for making exploding drones in Mexico." Fox 5 KVVU-TV. 24 April 2021, https://www.fox5vegas.com/news/us_world_news/2-to-stand-trial-for-making-exploding-drones-in-mexico/article_afcf0ee5-c4bc-53c9-a4e8-c0eafd61a381.html.

Associated Press, "Mexican drug cartels use exploding drones to attack police, soldiers." *El Paso Press*. 26 April 2021, https://www.elpasotimes.com/story/news/crime/2021/04/26/mexican-drug-cartels-use-explosive-drones-attack-police-soldiers/7384039002/.

B

Nelson Balido, "Nelson Balido: Mexican cartels patrol border with drones – and U.S. has no response." *Fox News Latino*, 19 February 2016 at http://latino.foxnews.com/latino/opinion/2016/02/19/nelson-balido-mexican-cartels-patrol-border-with-drones-and-us-has-no-response/.

Andrés Becerril, "Narcos copian bombas de FARC; Cisen alerta de explosivos tipo 'papa.'" *Excelsior.* 21 July 2017, http://www.excelsior.com.mx/nacional/2017/07/21/1176937.

Sarah Berger, "Mexico Drug Trafficking: Drone Carries 28 Pounds of Heroin Across Border To US." *International Business Times*, 13 August 2015 at http://www.ibtimes.com/mexico-drug-trafficking-drone-carries-28-pounds-heroin-across-border-us-2051941

Said Betanzos, "Cayó 'narcodrón' cerca de garita," *El Mexicano.* 22 January 2015 at http://www.el-mexicano.com.mx/informacion/noticias/1/3/estatal/2015/01/22/819263/cayo-narcodron-cerca-de-garita.

Robert J. Bunker, *Terrorist and Insurgent Unmanned Aerial Vehicles: Use, Potentials, and Military Implications.* Carlisle Barracks: Strategic Studies Institute, US Army War College, 2015, https://publications.armywarcollege.edu/pubs/2238.pdf.

Robert J. Bunker and John P. Sullivan, Eds. *Illicit Tactical Progress: Mexican Cartel Tactical Notes 2013-2020.* (A Small Wars Journal-El Centro Anthology.) Bloomington: Xlibris, 2021, https://www.amazon.com/Illicit-Tactical-Progress-Mexican-2013-2020/dp/1664180516.

Robert J. Bunker, John P. Sullivan, David A. Kuhn, "Use of Weaponized Consumer Drones in Mexican Crime War." *Counter-IED Report.* Winter 2020-2021, pp. 69-77, https://www.academia.edu/45000611/Use_of_Weaponized_Consumer_Drones_in_Mexican_Crime_War.

Jorge Butrón, "Caen en Puebla pioneros en uso de drones con explosivos." *La Razón*. 24 April 2021, https://www.razon.com.mx/estados/caen-puebla-pioneros-utilizar-drones-explosivos-432129.

C

"Caen cuatro hombres que transportaban un dron acoplado a un artefacto explosivo en Guanajuato." *Proceso*. 20 October 2017, http://www.proceso.com.mx/508237/caen-cuatro-hombres-transportaban-dron-acoplado-a-artefacto-explosivo-en-guanajuato/amp.

"Captan en video emboscada del CJNG contra autodefensas y reporteros." *El Universal*. 17 December 2020, https://www.eluniversal.com.mx/estados/captan-en-video-emboscada-del-cjng-contra-autodefensas-y-reporteros-0.

"Cártel del Tepalcatepec denunció incursión del CJNG con supuestos drones." *Infobae*. 5 May 2021, https://www.infobae.com/america/mexico/2021/05/05/cartel-del-tepalcatepec-denuncio-incursion-del-cjng-con-supuestos-drones/.

"'La casa baleada', fortaleza del CJNG desde la que mantienen una cruda batalla contra las autodefensas de Michoacán." *Infobae*. 18 December 2020, https://www.infobae.com/america/mexico/2020/12/18/la-casa-baleada-fortaleza-del-cjng-desde-la-que-mantienen-una-cruda-batalla-contra-las-autodefensas-de-michoacan/.

Rafa Fernandez De Castro, "Meth from Heaven? Narco drone falls out of Tijuana sky.' *Fusion,* 22 January 2015 at http://fusion.net/story/39975/meth-from-heaven-narco-drone-falls-out-of-tijuana-sky/.

Marguerite Cawley, "Drone Use in Latin America: Dangers and Opportunities." *InsightCrime*. 18 April 2014, http://www.insightcrime.org/news-analysis/drone-use-in-latin-america-dangers-and-opportunities.

Santiago Cepeda, "El peligroso mundo de las 'papas bomba'." *Revista DonJuan*. 18 de Abril de 2012, http://m.revistadonjuan.com/historias/el-peligroso-mundo-de-las-papas-bomba+articulo+11596882.

"CJNG ataca con drones cargados de explosivos comunidad de Tepalcatepec. Michoacán." *Político MX*. 4 May 2021, https://politico.mx/minuta-politica/minuta-politica-estados/cjng-ataca-con-drones-cargados-de-explosivos-comunidad-de-tepalcatepec-michoacán/.

"CJNG ataca con drones comunidades de Tepalcatepec, Michoacán." *El Universal*. 5 May 2021, https://www.eluniversal.com.mx/estados/cjng-ataca-con-drones-comunidades-de-tepalcatepec-michoacan.

"CJNG ataca con narcodrones; autodefensas de Tepalcatepec detallan su uso." *Televisa.NEWS*, 18 August 2020, https://noticieros.televisa.com/ultimas-noticias/cjng-ataca-con-narcodrones-autodefensas-de-tepalcatepec-michoacan-detallan-su-uso/.

"CJNG ataca con narcodrones; autodefensas de Tepalcatepec detallan su uso." *Noticeros Televisa*. 18 August 2020, https://noticieros.televisa.com/ultimas-noticias/cjng-ataca-con-narcodrones-autodefensas-de-tepalcatepec-michoacan-detallan-su-uso/.

"CJNG ataca a reporteros que acompañaban a autodefensas de Michoacán." *Vanguardia*. 19 December 2020, https://vanguardia.com.mx/articulo/cjng-ataca-reporteros-que-acompanaban-autodefensas-de-michoacan.

"#CJNG EMBOSCA A AUTODEFENSAS Y REPORTEROS EN #LABOCANDA VIDEO." *Valor Por Tamaulipas*. 18 December 2020, https://www.valorportamaulipas.info/2020/12/cjng-embosca-autodefensas-y-reporteros.html.

"El CJNG explotó un dron durante visita a Michoacán del Embajador del Vaticano; hirieron a 2 policías," *Sin Embargo*. 20 April 2021, https://www.sinembargo.mx/20-04-2021/3965563.

"CJNG 'trabaja desde el aire': utiliza drones para vigilar en Guanajuato." *La Silla Rota*. 21 April 2021, https://guanajuato.lasillarota.com/estados/cjng-trabaja-desde-el-aire-utiliza-drones-para-vigilar-en-guanajuato/509670.

"CJNG utiliza explosivos tipo 'papa' similares a las de las FARC." *El Debate*. 21 July 2017, https://www.debate.com.mx/mexico/CJNG-utiliza-explosivos-tipo-papa-similares-a-las-de-las-FARC-20170721-0066.html.

"CJNG usa drones con explosivos C4 y balines como forma de ataque." *El Universal*. 18 August 2020, https://www.eluniversal.com.mx/nacion/cjng-usa-drones-con-explosivos-c4-y-balines-como-forma-de-ataque.

"Comunicado FGR 148/21. FGR obtiene vinculación a proceso para dos hombres detenidos, uno en Puebla y otro en Morelos." Mexico City: Fiscalía General de la República (FGR). 24 April 2021, https://www.gob.mx/fgr/prensa/comunicado-fgr-148-21-fgr-obtiene-vinculacion-a-proceso-para-dos-hombres-detenidos-uno-en-puebla-y-otro-en-morelos.

"Con drones disparan y lanzan granadas contra policías en Aguililla, Michoacán." *Animal Politico*. 22 April 2021, https://www.animalpolitico.com/2021/04/drones-lanzan-granadas-policias-aguililla-michoacan/.

"Con drones envían granadas a casa de Sosa Olachea." *Zeta*. 10 July 2018, http://zetatijuana.com/2018/07/con-drones-envian-granadas-a-casa-de-sosa-olachea/.

"#ConferenciaPresidente | Miércoles 21 de abril de 2021." YouTube. 21 April 2021, https://www.youtube.com/watch?v=v9-b8EJhhIY.

"Con Granadas Transportadas En Un Dron Atacan Casa De Secretario De Seguridad Pública De BC." *Reporte Indigo*. 10 July 2018, https://www.reporteindigo.com/reporte/granadas-transportadas-en-dron-atacan-casa-secretario-seguridad-publica-bc/.

Marco A Coronel, "El #CártelJaliscoNuevaGeneración convirtió La Bocanda, Michoacán, en su base de operaciones." *Twitter.* 17 December 2020, https://twitter.com/marcocoronel/status/13397 87765513396226?s=20.

Counter-Unmanned Aircraft System Techniques. ATP 3-01.81. Washington, DC: Headquarters US Army, https://rdl.train.army. mil/catalog-ws/view/100.ATSC/9B8B46D7-719C-4E15-A8FE-9F2C1E278B88-1492434973380/atp3_01x81.pdf.

Rob Crilly, "Meth-laden drone crashes near US-Mexico border." *The Telegraph,* 22 January 2015 at http://www.telegraph.co.uk/news/ worldnews/centralamericaandthecaribbean/mexico/11361825/Metha-drone-crashes-near-Mexican-border.html.

Jeanna Cullinan, "Mexico to Use Drones Against Drug Cartels." *InSight Crime.* 17 November 2011, https://www.insightcrime.org/news/brief/ mexico-to-use-drones-against-drug-cartels/.

D

Chris Dalby, "How Mexico's Cartels Have Learned Military Tactics." *InSight Crime.* 2 September 2021, https://insightcrime.org/news/ how-mexicos-cartel-have-learned-military-tactics/.

José Jiménez Díaz, "No detona dron con granadas en casa de Sosa Olachea." *El Mexicano.* 10 July 2018, http://www.el-mexicano. com.mx/informacion/noticias/1/3/estatal/2018/07/10/1052433/ no-detona-dron-con-granadas-en-casa-de-sosa-olachea.

"Desconocidos pegan granadas a dron y las hacen caer en casa del Secretario de Seguridad Pública de BC." *Sin Embargo.* 10 July 2018, http://www.sinembargo.mx/10-07-2018/3440637.

"Descubren un 'narcodron' en Colombia; enviaba cocaína a Panamá (Authorities Discover Narco-Drone in Colombia with Cocaine

Shipment Headed to Panama)," *Excelsior*. 15 November 2016. http://www.seguridadenamerica.com.mx/noticias/de-consulta/secciones-revist-seguridad-en-america/noticias-de-control-de-acceso/24684-descubren-un-narcodron-en-colombia-enviaba-cocaina-a-panama.

"Destruyen drones con explosivos que pertenecerían a disidencias en Nariño." *BLU Radio*.18 September 2019, https://www.bluradio.com/nacion/destruyen-drones-con-explosivos-que-pertenecerian-a-disidencias-en-narino.

"'Dron bomba' listo para detonar a distancia." *AM*. 20 October 2017, https://www.am.com.mx/2017/10/20/sucesos/dron-bomba-listo-para-detonar-a-distancia-385808.

"Drones bomba: la nueva arma de las disidencias en Tumaco." *Semana*. 19 September 2019, https://www.semana.com/nacion/articulo/drones-bomba-la-nueva-arma-de-las-disidencias-en-tumaco/632441/.

"Drones con explosivos, la más reciente arma del CJNG para atacar desde el aire." *Infobae*. 14 August 2020, https://www.infobae.com/america/mexico/2020/08/15/drones-con-explosivos-la-mas-reciente-arma-del-cjng-para-atacar-desde-el-aire/.

"Drone Drug Smuggler Gets 12-Year Sentence." *CBP Newsroom*. 31 July 2018, https://www.cbp.gov/newsroom/national-media-release/drone-drug-smuggler-gets-12-year-sentence.

"Drones explosivos de Aguililla, funcionaron mal: Sedena." *MoreliActiva*. 21 April 2021, https://moreliactiva.com/drones-explosivos-de-aguililla-funcionaron-mal-sedena/.

"Dron explosivo: Último artefacto del crimen organizado en México." *HispanTV*. 21 October 2017, http://www.hispantv.com/noticias/mexico/357219/incautan-dron-crimen-organizado-violencia.

"Drone with grenades falls into official's residence." *Imperial Valley Press*. 11 July 2018, https://www.pressreader.com/usa/imperial-valley-press/20180711/281582356394974.

E

"Ejército halla dos drones cargados con explosivos en zona rural de Tumaco, Nariño." *El País*. 18 September 2019, https://www.elpais.com.co/judicial/ejercito-halla-dos-drones-cargados-con-explosivos-en-zona-rural-de-tumaco-narino.html.

"Ejército desactivó dos drones cargados con explosivos en Tumaco." *El Tiempo*. 18 September 2019, https://www.eltiempo.com/justicia/investigacion/ejercito-desactivo-dos-drones-cargados-con-explosivos-en-tumaco-413816.

Héctor Estepa, "Del narco al 'coyote': el mundo del crimen latinoamericano reinventa el uso del dron." *El Confidencial*. 1 January 2020, https://www.elconfidencial.com/mundo/2020-01-01/del-narco-al-coyote-los-nuevos-usos-del-dron-en-el-mundo-del-crimen-latino americano_2391035/.

Jocelyn Estrada, "Ataque contra policías en Aguililla fue con explosivos plásticos: SSP de Michoacán." *Milenio*. 21 April 2021, https://www.milenio.com/estados/ataque-policias-aguililla-explosivos-plasticos-ssp.

F

"Fabrican narcos sus propios drones, alerta la DEA (The DEA Reports that Narcos are Building their Own Drones)," *La Nacion*, 09 June 2014. http://archivo.eluniversal.com.mx/nacion-mexico/2014/carteles-fabrican-narcodrones-trasiego-eu-1022274.html.

Vanda Felbab-Brown, "Drugs and Drones: The Criminal Empire Strikes Back." *Remote Control Project Blog*, 24 February 2016 at http://

remotecontrolprojectblog.org/2016/02/24/drugs-and-drones-the-crime-empire-strikes-back/.

"FGR secures bombs and drones for terrorism, in Cholula." *EN24*. 4 May 2020, https://en24.news/en/2020/05/fgr-secures-bombs-and-drones-for-terrorism-in-cholulahtml.

G

Octavio Ortiz García, "Drones explosivos de Aguililla, funcionaron mal: Sedena." *MoreliActiva*. 21 April 2021, https://moreliactiva.com/drones-explosivos-de-aguililla-funcionaron-mal-sedena/.

Camilo Mejia Giraldo, "Mexico's Cartels Building Custom-Made Narco Drones: DEA." *Insight Crime*. 11 July 2014, http://www.insightcrime.org/news-briefs/mexicos-cartels-building-custom-made-narco-drones-dea.

Doris Gómora, "Fabrican narcos sus propios drones, altera la DEA." *El Universal*. 9 July 2014, https://archivo.eluniversal.com.mx/nacion-mexico/2014/carteles-fabrican-narcodrones-trasiego-eu-1022274.html.

Ángel F. González, "Señala SSPE a cárteles en atentado a secretario." *Frontera.info*. 11 July 2018, http://www.frontera.info/EdicionEnLinea/Notas/Noticias/11072018/1355958-Senala-SSPE-a-carteles-en-atentado-a-secretario.html.

Juan Manuel González, "Con drones, CJNG busca erradicar a rivales en Tierra Caliente." *La Silla Rota*. 12 August 2020, https://lasillarota.com/estados/con-drones-cjng-busca-erradicar-a-rivales-en-tierra-caliente-michoacan-drones-cjng-mencho/423494.

Ronan Graham, "US Cracks Down on Drug Smugglers in Ultralight Planes." *InSight Crime*. 15 December 2011, https://www.insightcrime.org/news/brief/us-cracks-down-on-drug-smugglers-in-ultralight-planes/.

H

David Hambling, "Mexican Cartel Injures Police Officers With Drone Bomb Attack." *Forbes*. 22 April 2021, https://www.forbes.com/sites/davidhambling/2021/04/22/mexican-cartel-injures-police-officers-with-drone-bomb-attack/?sh=352d0068127a.

Gina Harkins, "Illicit drone flights surge along U.S.-Mexico border as smugglers hunt for soft spots." *Washington Post*. 24 June 2018, https://www.washingtonpost.com/world/national-security/illicit-drone-flights-surge-along-us-mexico-border-as-smugglers-hunt-for-soft-spots/2018/06/24/ea353d2a-70aa-11e8-bd50-b80389a4e569_story.html?utm_term=.0992752ea810.

Ana Vanessa Herrero and Nicholas Casey, "Venezuelan President Targeted by Drone Attack, Officials Say." *New York Times*. 4 August 2018, https://www.nytimes.com/2018/08/04/world/americas/venezuelan-president-targeted-in-attack-attempt-minister-says.html.

Jan-Albert Hootsen, "Inside Mexico's Drone Wars." *Voacative*, 6 January 2014 at http://www.vocativ.com/world/mexico-world/inside-mexicos-drone-wars/.

I

"Interceptan ¡dron bomba!" *AM*. 20 October 2017, https://www.am.com.mx/2017/10/20/leon/sucesos/interceptan-dron-bomba-385781.

J

"Jalisco cartel adopts new tactic: drones armed with C-4 explosive." *Mexico News Daily*. 18 August 2020, https://mexiconewsdaily.com/news/jalisco-cartel-adopts-new-tactic-drones-armed-with-c-4-explosive/.

Katie Jones, "How Organized Crime Networks Are Using Drones to Their Advantage." *InSight Crime.* 29 September 2020, https://www.insightcrime.org/news/brief/drones-narcotrafficking-surveillance/.

K

Zachary Kallenborn. "The Era of the Drone Swarm is Coming, and We need to be Ready for It." *Modern War Institute.* 25 October 2018, https://mwi.usma.edu/era-drone-swarm-coming-need-ready/.

Zachary Kallenborn and Phillip C. Bleek, "Drones of Mass Destruction: Drone Swarms and the Future of Nuclear, Chemical, Biological Weapons." *War on the Rocks.* 14 February 2019, https://warontherocks.com/2019/02/drones-of-mass-destruction-drone-swarms-and-the-future-of-nuclear-chemical-and-biological-weapons/.

Alma Keshavarz and Robert J. Bunker, "Weaponized Drone Linked to Organized Crime in Mexico." *OE Watch.* Vol. 7., Iss. 11. December 2017: p. 33.

Alma Keshavarz and Robert J. Bunker, "Weaponized Drones Target Baja California Secretary of Public Security's Residence." *OE Watch.* Vol. 8., Iss. 8., August 2018: p. 71.

Jeremy Kryt, "Game of Drones: Mexico's Cartels Have a Deadly New Weapon." *The Daily Beast.* 12 November 2017, https://www.thedailybeast.com/game-of-drones-mexicos-cartels-have-a-deadly-new-weapon.

L

"Liberan policías de Michoacán bloqueo carretero entre Aguililla y Apatzingán; los atacan con drones." *Aristegui Noticias.* 20 April 2021, https://aristeguinoticias.com/2004/mexico/libera-ssp-michoacan-bloqueo-de-carretera-entre-aguililla-y-apatzingan/.

M

"Maduro asegura que involucrados en atentado en su contra se entrenaban con personal colombiano (Maduro said that those involved in the assassination attempt trained with Colombian personnel)," *El Universal*, 7 August 2018. http://www.eluniversal.com/politica/17050/maduro-asegura-que-involucrados-en-atentado-en-su-contra-se-entrenaban-con-personal-colombiano.

Martha Maguire, "US Warns of Rise in Secret Tunnels Under Mexico Border." *InSight Crime*. 16 June 2011, https://www.insightcrime.org/news/brief/us-warns-of-rise-in-secret-tunnels-under-mexico-border/.

"Mexico cartel used explosive drones to attack police." *BBC News*. 21 April 2021, https://www.bbc.com/news/world-latin-america-56814501.

Jorge Monroy, "Uso de drones con explosivos, actos terroristas: SSP de Michoacán." *El Economista*. 21 April 2021, https://www.eleconomista.com.mx/politica/Uso-de-drones-con-explosivos-actos-terroristas-SSP-de-Michoacan-20210421-0092.html.

Rubén Mosso, "FGR asegura explosivos y drones en Puebla; investiga terrorismo." *Milenio*. 3 May 2020, https://www.milenio.com/policia/inicia-fgr-pesquisa-delito-finalidad-cometer-terrorismo.

Rubén Mosso, "Indaga FGR terrorismo tras hallar C4 en Puebla." *Milenio*. 4 May 2020, https://www.milenio.com/policia/inicia-fgr-pesquisa-delito-finalidad-cometer-terrorismo.

N

"Narcos, disidencias y ELN estarían utilizando drones para espiar a la fuerza pública." *BLU Radio*. 19 September 2019, https://www.bluradio.com/nacion/narcos-disidencias-y-eln-estarian-utilizando-drones-para-espiar-a-la-fuerza-publica.

"Los narcodrones de la frontera." *Telemundo Local*, 3 February 2015 at http://www.telemundo51.com/noticias/Los-narcodrones-de-la-frontera-narcotrafico-nogales-arizona-mexico-eeuu-290686221.html.

"'Narco drones' puts all U.S. border efforts in question." *The American Post*. 2010, https://www.theamericaspostes.com/2338/narco-drones-puts-all-u-s-border-efforts-in-question/.

"'Narcodrones', la nueva técnica de los cárteles mexicanos." *El Comercio*, 16 July 2014 at http://elcomercio.pe/mundo/latinoamerica/narcodrones-nueva-tecnica-carteles-mexicanos-noticia-1743520.

"Narco envía droga a EU… en drones (Drug Traffickers Use Drones to Ship Drugs to the United States)," *Ejecentral*, 17 August 2014. http://www.eluniversal.com.mx/nacion-mexico/2014/carteles-fabrican-narcodrones-trasiego-eu-1022274.html.

"Narcotraficantes envian cocaina a Panama con drones: Policia de Colombia (Colombian Drug Traffickers Send Cocaine to Panama with Drones)." *La Prensa*, 17 November 2016. http://www.prensa.com/mundo/Narcotraficantes-enviando-Panama-Policia-Colombia_0_4622537754.html.

"Neutralizados y destruidos dos drones cargados de explosivos." *La Seguridad es de Todos*. 19 September 2019, https://www.ejercito.mil.co/index.php?idcategoria=467681.

"Niegan Los CJNG El Uso De Drones Con Explosivos Video." *Valor Por Tamaulipas*. 25 April 2021, https://www.valorportamaulipas.info/2021/04/niegan-los-cjng-el-uso-de-drones-con.html.

O

Andrew O'Reilly, "DEA: Narco-drones not major smuggling concern, but could help set up attacks on agents." *Fox News Latino*, 22 January 2015 at http://latino.foxnews.com/latino/news/2015/01/22/

dea-narco-drones-not-major-smuggling-concern-but-could-help-set-up-attacks-on/.

P

"Policía revela 'el hormigueo', nueva modalidad del Clan Úsuga para sacar droga del país (Police Reveal New Micro-Trafficking Scheme Utilized by Clan Úsuga to Move Drugs from Colombia to Panama)," *Noticias CMI*, 10 October 2016. http://www.cmi.com.co/policia-revela-el-hormigueo-nueva-modalidad-del-clan-Úsuga-para-sacar-droga-del-pais.

Sol Prendido, "Michoacán, Mexico: CJNG Denies Involvement in Drone Bomb Attack." *Borderland Beat*. 25 April 2021, http://www.borderlandbeat.com/2021/04/michoacan-mexico-cjng-fuerzas.html.

"Procesaron a dos fabricantes de drones con explosivos que trabajaban para el 'Marro', ex líder del CSRL." *Infobae*. 24 April 2021, https://www.infobae.com/america/mexico/2021/04/25/procesaron-a-dos-fabricantes-de-drones-con-explosivos-que-trabajaban-para-el-marro-ex-lider-del-csrl/.

Q

"Qué hay detrás de los supuestos drones con explosivos "asegurados" en Tepalcatepec." *Noventa Grados (90º)*. 14 August 2020, http://www.noventagrados.com.mx/seguridad/que-hay-detras-de-los-supuestos-drones-con-explosivos-asegurados-en-tepalcatepec.htm.

"¿Qué se sabe del atentado contra el presidente Nicolás Maduro? (What is known about the attack against President Nicolas Maduro?)." *TeleSUR*. 5 August 2018, https://www.telesurtv.net/multimedia/maduro-claves-atentado-venezuela-20180805-0035.html.

"President Nicolas Maduro Reveals Evidence For Drone Attack." *TeleSUR*. 7 August 2018, https://www.youtube.com/watch?v=TPD3WSW6wAs.

R

"Reportan drones con granadas; uno descendió en la casa del titular e la SSP en BC." *Proceso*. 10 July 2018, https://www.proceso.com.mx/542389/reportan-drones-con-granadas-uno-descendio-en-la-casa-del-titular-de-la-ssp-en-bc.

S

Arturo Salinas y Manuel Ocaño, "FOTOGALERÍA: Cae dron que transportaba droga en Tijuana." *Excelsior*. 22 January 2015, http://www.excelsior.com.mx/nacional/2015/01/22/1003922.

Aaron R. Schmersahl, *Fifty Feet Above the Wall: Cartel Drones in the U.S.-Mexico Border Zone Airspace, and What to Do About Them.* Monterey, CA: Naval Post Graduate School, March 2018, https://www.hsdl.org/?view&did=811367.

"La Sedena confirma que el CJNG ha usado drones con explosivos en Michoacán y en Guanajuato." *Sin Embargo*. 21 April 2021, https://www.sinembargo.mx/21-04-2021/3965937.

Noé Cruz Serrano, "Ejército va por sistema antidrones de cárteles del narco." *El Universal*. 21 September 2020, https://www.eluniversal.com.mx/nacion/ejercito-va-por-sistema-antidrones-de-carteles-del-narco.

"La Sedena confirma que el CJNG ha usado drones con explosivos en Michoacán y en Guanajuato." *Sin Embargo*. 21 April 2021, https://www.sinembargo.mx/21-04-2021/3965937.

Scott Smith and Christine Armario, "Venezuela's Maduro: Drone attack was attempt to kill him." *Washington Post*. 4 August 2018, https://www.washingtonpost.com/world/the_americas/venezuelan-government-drone-strikes-targeted-maduro/2018/08/04/01034d9a-9846-11e8-818b-e9b7348cd87d_story.html?utm_term=.b55612dac292.

Mark Stevenson, "México: Cárteles atacan con drones cargados de explosivos." *Los Angeles Times*. 21 April 2021, https://www.latimes.com/espanol/mexico/articulo/2021-04-21/mexico-carteles-atacan-con-drones-cargados-de-explosivos.

John P. Sullivan "From Drug Wars to Criminal Insurgency: Mexican Cartels, Criminal Enclaves and Criminal Insurgency in Mexico and Central America. Implications for Global Security." *Working Paper Nº 9*, April 2012. Paris: Fondation Maison des sciences de l'homme, https://halshs.archives-ouvertes.fr/halshs-00694083/document.

John P. Sullivan, "Mexican Cartel Adaptation and Innovation." *OODA Loop*. 27 January 2020, available at https://www.academia.edu/41754168/Mexican_Cartel_Adaptation_and_Innovation.

John P. Sullivan, José de Arimatéia da Cruz, and Robert J. Bunker, "Third Generation Gangs Strategic Note No. 42: Brazilian Gangs Utilize Human Shields, Explosives, and Drones in a New 'Cangaço' Style Urban Bank Raid in Araçatuba, São Paulo." *Small Wars Journal*. 5 September 2021, https://smallwarsjournal.com/jrnl/art/third-generation-gangs-strategic-note-no-42-brazilian-gangs-utilize-human-shields.

T

"Tapan unos, el crimen troza otros caminos, pero la Policía Michoacán seguirá en Aguililla, asegura Israel Patrón." *La Voz Michoacán*. 21 April 2021, https://www.lavozdemichoacan.com.mx/michoacan/tapan-unos-el-crimen-troza-otros-caminos-pero-la-policia-michoacan-seguira-en-aguililla-asegura-israel-patron/.

David Teiner, "Cartel-Related Violence in Mexico as Narco-Terrorism or Criminal Insurgency: A Literature Review." *Perspectives on Terrorism*. Vol. 14, no. 4. August 2020: pp. 83-98, available at https://www.jstor.org/stable/26927665.

"¿Tomar video? Dron con fines terroristas." REDACCIÓN. *Am.* 21 October 2017, https://www.am.com.mx/2017/10/21/local/tomar-video-dron-con-fines-terroristas--386171.

"2 to stand trial for making exploding drones in Mexico." *Mercury News.* 24 April 2021, https://www.mercurynews.com/2021/04/24/2-to-stand-trial-for-making-exploding-drones-in-mexico/.

U

Alexandra Ulmer and Vivian Sequera, "Venezuela's Maduro says drone blast was bid to kill him, blames Colombia." Reuters Venezuela. 4 August 2018, https://www.reuters.com/article/us-venezuela-politics/venezuelas-maduro-target-of-drone-attack-but-unharmed-government-idUSKBN1KP0SA.

V

Rubens Valente, "Drones assustam indígenas em terra alvo de madeireiros e fazendeiros no MA." *UOL.* 14 September 2020, https://noticias.uol.com.br/colunas/rubens-valente/2020/09/14/drones-invasao-terras-indigenas-maranhao.htm.

"Viajaban en auto robado: les hallan un dron y explosivos." *El Debate.* 21 October 2017, https://www.debate.com.mx/mexico/Viajaban-en-auto-robado-les-hallan-un-dron-y-explosivos-20171021-0007.html.

Edmundo Velázquez, "FGR investiga si el Cártel de Santa Rosa de Lima maquilaba artefactos explosivos en San Pedro Cholula." *Periodico Central.* 1 May 2020, https://www.periodicocentral.mx/2020/pagina-negra/delincuencia/item/9079-fgr-investiga-si-el-cartel-de-santa-rosa-de-lima-maquilaba-artefactos-explosivos-en-san-pedro-cholula.

W

Jacob Ware, "Terrorist Groups, Artificial Intelligence, and Killer Drones." *War on the Rocks.* 24 September 2019, https://warontherocks.com/2019/09/terrorist-groups-artificial-intelligence-and-killer-drones/.

Illicit Tactical Progress
Mexican Cartel Tactical Notes 2013-2020
A Small Wars Journal-El Centro Anthology

Robert J. Bunker and John P. Sullivan, Editors

Publication Date: 26 July 2021

Format: Softcover
Publisher: Xlibris (Small Wars Foundation)
Page Count: 360
ISBN: 9781664180512

This book is an eye-opener that may be an appalling representation of current events in Mexico, but it is based on factual reports of the strength, manner, and frequency of the cartel violence that occurs every day in Mexico. ... As long as the cartels continue to keep their wars inside Mexico and as long as Mexico does not ask for US help, the status quo will continue, and we will see this level and scope of violence incrementally increase in that nation.

Gary J. Hale, DEA (Ret.)

Back Cover Images: *Top*; 3DR Solo Quadcopter with IED and remote detonation switch (side view). 20 October 2017. Source; *Policía Federal* (PF), Guanajunto, Mexico [For Public Distribution//No Restrictions on Use]. *Bottom Left*; Drone suspected of hauling 12 packages of meth two miles west of the San Ysidro Port of Entry.18 August 2017 (Release Date). U.S. Customs and Border Protection [For Public Distribution// No Restrictions on Use]. *Bottom Center*; Photo of "Spreading Wings 900" Drone (*Dron*) recovered by Tijuana Municipal Police on 20 January 2015 near San Ysidro border crossing in the Zona del Río. The drone was carrying about six pounds of methamphetamine (*cristal*) and had crashed in a parking lot. Source: *Secretaria de Seguridad Pública Municipal Tijuana*, Mexico. *Bottom Right*; Tarot Ironman Drone hard landing in a courtyard with tennis court in the background. At the residence of Gerardo Sosa Olachea, the public safety secretary/ *Secretario de Seguridad Pública Estalal* (SSPE) of Baja California, in *colonia* Los Laureles in Tecate. 10 July 2018. Courtesy of *Zeta*. [For Public Distribution//Used with Permission]

Printed in the United States
by Baker & Taylor Publisher Services